AF521694

TO A YOUNG BIRD ARTIST

LETTERS FROM

LOUIS AGASSIZ FUERTES

TO GEORGE MIKSCH SUTTON

TO A YOUNG BIRD ARTIST

LETTERS FROM LOUIS AGASSIZ FUERTES TO GEORGE MIKSCH SUTTON

Commentary by
George Miksch Sutton

UNIVERSITY OF OKLAHOMA PRESS : NORMAN

Library of Congress Cataloging in Publication Data
Fuertes, Louis Agassiz, 1874–1927.
To a young bird artist.
1. Fuertes, Louis Agassiz, 1874–1927. 2. Sutton, George Miksch, 1898– 3. Ornithologists—United States—Correspondence. I. Sutton, George Miksch, 1898– II. Title.
QL31.F83A37 598′.092′4 [B] 79–4754
ISBN 0–8061–1589–0

For Mary and Sumner

PREFACE

My decision to publish the letters that renowned bird artist Louis Agassiz Fuertes wrote me between 1915 and 1927, the year of his tragic death in an automobile accident, has been prompted chiefly by realization that they contain sound advice for aspiring young bird artists, but also because there are in them profoundly philosophical statements that apply to anyone, whatever his talents or field of endeavor.

Fuertes was not obliged to take me under his wing. No one told him that he should pay attention to this youngster who clamored so earnestly for attention in 1915. He was a busy man. He had deadlines to meet, a family to provide for. Yet cheerfully he took me on, almost as if he had been waiting for this opportunity to help someone even as he had been helped by others when he was younger. Never was he impatient with me. Almost invariably he called attention first to what was good about my work, then to its shortcomings.

I have no false notions about my standing today as a portraitist of birds. I have made many bird pictures, some good, some less than good. Many of them, even several of the poorer ones, have been reproduced, so I have won for myself something of a reputation. But I am publishing these letters not because I consider myself shining proof that they produced results, but because I know that they helped me. Some of this help was of a very practical sort; some of it was inspirational.

All but two of the letters published here were written in longhand. Two were typewritten. One longhand letter, a lengthy one, was written with pencil, the rest with ink. Certain words in them have always been a bit hard to decipher. Where Fuertes used initials or other means of abbreviation, I have taken the liberty of filling in so as to make the meaning readily clear. Except where otherwise indicated, the letters were written at, and mailed from, Ithaca, New York.

Most of the letters that I wrote to Fuertes are on file in the Department of Manuscripts and University Archives at Cornell University. Xerox copies of these have been sent to me by Fuertes's daughter, Mary (Mrs. Damon Boynton), who has been good enough, too, to decipher one or two of the difficult words in her father's letters and to answer questions about activities of the Fuertes family and their friends at Sheldrake Point.

Mary adored her father. With me she felt that he was phenomenal—a human being with powers beyond comprehension. Writing in a recent letter, she so well sums up that wonderful man's attributes and dynamic, outgoing personality that I must quote a bit from it here: "But the amount of correspondence! I don't see how Daddy had time to do more than *open* his mail, let alone read it, let alone answer it, let alone paint, let alone lecture, let alone take care of all the neighborhood and community chores and pleasures. What stamina!"

In Mary's scholarly book, *Louis Agassiz Fuertes, His Life Briefly Told and His Correspondence Edited* (1956, Oxford University Press, New York, 317 pp.), she quoted two of the letters that I am publishing here—the first two, written February 15 and March 30, 1915—and a few paragraphs from the long letter of August 3, 1917. Only one other part of this book of mine, the part dealing with my stay with the Fuertes family at Sheldrake Point in the summer of 1916, has appeared elsewhere in print. It was published, not quite verbatim, under the title "Fuertes Remembered" (Sutton, 1974, *Audubon Magazine*, 76: 58–67).

At the very height of his career, Louis Agassiz Fuertes found time for me. Never for a moment did he let me feel that I was a burden. I marvel, I simply marvel, every time I think of the trouble he went to when he put up that tent for me near his house on the shore of Lake Cayuga. And that was only one of the many things he did for me.

I wish to thank Kenneth C. Parkes, Curator of Birds at the Carnegie Museum of Natural History in Pittsburgh, and Olin Sewall Pettingill Jr., formerly Director of the Laboratory of Ornithology at Cornell University, and now a resident of Wayne, Maine, for their timely assistance in the preparation of this book.

GEORGE MIKSCH SUTTON

TO A YOUNG BIRD ARTIST

LETTERS FROM

LOUIS AGASSIZ FUERTES

TO GEORGE MIKSCH SUTTON

THE FUERTES LETTERS

It was in Eugene, Oregon, in the Biology Laboratory on an upper floor of a building on the University of Oregon campus, that I first became aware of the bird drawings of Louis Agassiz Fuertes. The reproductions, though not in color, were lifelike beyond belief. They held me spellbound. The artist's signature fascinated me, though when I tried to tell my parents about it I could not pronounce either the middle name or the last. I had to spell the words out as best I could, from memory. The year was 1908. I was ten years old.

All my life I had been interested in birds. In my younger days, while the family was living in Nebraska, I had made crayon drawings of hawks, ducks, redbirds, bluebirds, and blackbirds, without bothering to ascertain what the real names of such birds might be. When my parents had given me a copy of Frank M. Chapman's *Bird Life,* I had discovered that kingbird was the proper name for the spunky little bird that all the neighbors called the bee martin; that bullbats were not bats at all, but nighthawks; and that the shiny blackbird with trough-shaped tail was actually a grackle. The many colorplates in *Bird Life* had been helpful. They had made clear what the birds looked like. But I had consulted them without being much excited by them.

In the Biology Laboratory at the University of Oregon there was a small collection of birdskins. How I learned of them I do not know, for my father was not on the university's faculty, so he could hardly have known where the Biology Laboratory was, or that there was a bird collection in it. I may have discovered the laboratory quite by chance; the building housing it had a feature that attracted all the small boys of the neighborhood—a smooth, highly polished rail leading steeply and without interruption from third floor down to first, atop the front stairway's balustrade. The door to the laboratory was only a few feet from that irresistible shiny rail.

Professor Albert R. Sweetser allowed me to study the birdskins. Most of the species were western, hence new to me, but a book titled *Handbook of Birds of the Western United States*, on a shelf close by, helped greatly. The author's name was Florence Merriam Bailey. The book had not a single colorplate. Most of the illustrations were small, but the several full-page halftones were the most exciting bird pictures I had ever seen. They had a powerful quality that—for want of a more explicit word—I called life. So very alive was the picture of the Steller's Jay that I turned to it again and again, astounded that the eye, as represented, seemed to twinkle, that the plumage seemed to be lifting and sinking as its wearer breathed. The picture was only a reproduction, to be sure; I was looking at paper, not a real bird. But somehow the real bird was there.

So often did I visit the laboratory to pore over the Bailey book and the birdskins that Professor Sweetser asked me to keep his collection in order for him. After class-use, it was sometimes a jumble, and I thoroughly enjoyed lining up all the thrushes, all the swallows, all the finches, side by side. Thus did I become a curator of sorts.

My interest in delineating birds took a serious turn. I had neither crayons nor paints, but I made a series of pencil drawings that I decided were good enough to keep. These I pasted together, end to end, making a roll of them. The bird figures were crude indeed, but they were much the best I had ever done. So determined was I to make clear through them that I was paying attention to detail that I overdid badly with one bird, giving its nearer leg two heels rather than one, with a short extra bone between heels.

As I look at these drawings today they seem less crude than funny. The names I gave my subjects, some misspelled, some fanciful, were funny too. There was, for example, a "Least Pheode." The influence of the Bailey book, with its lovely Fuertes plates, was evident enough: otherwise how could I have given the one woodpecker of the series the name Gairdner's Woodpecker—a name straight out of Bailey (see p. 203). The drawings are folded now, not rolled. But they are pasted together end to end, just as they were seventy years ago. I enjoy looking at them—and I smile as I look.

When, in 1909, the Sutton family moved to Illinois, I continued to study birds and to draw them. A picture that I made at school of an oriole and its hanging nest must have been considered good by my art teacher, for it was reproduced in a state annual of some sort. Needless to say, I drew that oriole from memory, not from a living model, or even a dead one, but it was a sincere bit of work. By this time I had learned of, and subscribed for, a folksy little magazine called *The Oölogist,* a sort of trade journal for egg collectors. Here there were very few bird drawings; but in a larger, more elegant magazine, a bi-monthly called *Bird-Lore,* were fine colorplates illustrating the plumages, molts, and migrations of American birds found north of Mexico. These plates, each showing several species, were so deeply appealing, so wonderfully satisfying, that they made my spine tingle. They were by Louis Agassiz Fuertes.

In the summer of 1911 the Sutton family moved to Fort Worth, Texas, where I fell in love with the prairie, began keeping a daily record of birds seen, started a collection of mounted birds, and made detailed drawings from freshly shot specimens. For a year or so we lived in a residential part of the city; then we moved to the vicinity of the Texas Christian University campus, an unwooded area in which there were very few houses. In the spring of 1913 I found a Roadrunner nest, in it eight young birds, one of which I took home with me. On April 22 I made a wash drawing of this captive—one of the very few drawings I had ever made directly from a living model. On May 16 I received from my parents a birthday present that I would treasure the rest of my life—my own copy of *Handbook of Birds of the Western United States.* How I adored that book! The picture of the Steller's

Jay roused my admiration and disbelief even as it had when I had first seen it: how could a mere drawing capture life as that drawing had? How could hardness of bill, scaliness of toes, softness of plumage, and sparkle of eye all be put on paper so convincingly?

So entertaining was my pet Roadrunner that I could not resist the urge to tell the world about it. In the September-October 1913 issue of *Bird-Lore,* in a section headed "From Young Observers," appeared my article. Its title: "A Pet Roadrunner."

When the winsome bird was able to fend for itself, I returned it to the wild. The following year, however, I reared two more Roadrunners, and one of these I drew soon after I had observed its odd behavior just after it had been frightened by a big toad. Neither of my Roadrunners would eat toads. They ate grasshoppers, cave crickets, mice, snakes, centipedes, tarantulas, scorpions, and lizards of all sorts—including the slow-moving ones that everybody called "horny frogs"—but they drew the line at toads. Small toads they looked at dubiously, as if remembering the taste of the last one they'd tried to eat. Large toads frightened them, causing them to rattle their bills sharply, lift their wings, and spread their tails wide. I kept the sketch; it was a by-product of my determination to make drawings that would look alive, even as the Fuertes drawings did.

Most of the drawings that I made in those days must have been imitative of Fuertes to some extent, but never did I copy his work outright, nor did I copy photographs.

In July of 1914, the peripatetic Suttons moved once more, this time to West Virginia. With us on the train went two Roadrunners and two Screech Owls. All four of these pets I liberated on the Bethany College campus, where my father was to teach. The owls I never saw again. But the Roadrunners hung around for weeks, begging for food from any passerby; finding grasshoppers in sunny places away from the woods; and catching House Sparrows at spots where trash was dumped. One of my pets was killed by a man who, having heard of my interest in birds, felt that he should obtain for me this strange one that ran up to him with mouth open and wings fluttering, begging for food. The other strayed farther and farther from the campus as summer waned, and by fall had disappeared.

In the January-February 1915 issue of *Bird-Lore* appeared my drawing of a pet Roadrunner "in an attitude of fright." At the front of this same issue was an exquisite colorplate of juncos by Louis Agassiz Fuertes. Publication of my drawing convinced me that somehow I had become a bird artist. It convinced me that I had a right to make myself known to Fuertes, to become personally acquainted with him. Perhaps, through knowing him, I could learn to impart life to my drawings.

I decided to write to Fuertes. I learned from something that appeared in *Bird-Lore* that he lived in Ithaca, New York. I wrote a letter. Indeed, I wrote it several times, each time discussing it with my parents, though I did not show it to them. They decided that under the circumstances—circumstances that at times must have made them somewhat frantic—a letter would be in order. It might seem presumptuous to Mr. Fuertes. It might bore him. In any event, he would not need to reply to it unless he wanted to.

In late January or early February I mailed the letter. Some of what I said in it (I kept no copy) is clear enough from the reply:

LOUIS AGASSIZ FUERTES
CORNELL HEIGHTS, ITHACA, N.Y.

Feb. 15, 1915

Mr. George Miksch Sutton,
Bethany, W. Va.

My dear Mr. Sutton—

Your very kind letter came to me out of a clear sky, and I cannot well tell you how much I was touched and gratified by your warm appreciation of my efforts to come at the truth in bird painting. Much praise from the uninitiated means little or nothing: a rare tribute like yours, from those who have met some of the difficulties & know them by their first names is an uplift and encouragement that means more than all the every day, inexpert praise in the world.

So far as I know, I shall be in or near Ithaca this summer, and if you should decide to come up, will be glad to help you in any way I can. My methods are not academic, and have been developed to suit the rather peculiar needs of the situation, and I might do you more harm than good.

In the meantime, if you have any study material, or pictures that you have done and would like criticism on, I would be glad if you would send them on, and I will write you more fully than I can do now, as to their promise, the field, etc. etc—all vital points if you are thinking of going seriously into it. Of course the field is limited, and must in a large measure be created by the one who is to fill it. When I started, about 20 years ago, there wasn't any field, but one developed, and now there is considerably more of a one than one man can fill, though there are not many who seem to combine the sympathy with nature, the specific knowledge of their subject, and the technical ability to paint, which are all necessary if their work is to stand up under the test of time and the inevitable competition that is ahead. Say what you will of Audubon—(much of what you did say is just, too) he was the first and only man whose bird drawings showed the faintest hint of anatomical study, or that the fresh bird was in hand when the work was done, and is so immeasurably ahead of anything up to his time or since, until the modern idea of drawing endlessly from life began to bear fruit, that its strength deserves all praise and honor, and its many weaknesses condonement, as they were the fruit of his training:- stilted, tight and unimaginative old David sticks out in the stiff landscape, the hard outline and the dull, lifeless shading, while the overpowering virility of A. him-

self is shown in the snappy, instantaneous attitudes, and dashing motion of his subjects. While there's much to criticize there is also much to learn, and much to admire, in studying the monumental classic that he left behind him. He made many errors, but he also left a living record that has been of inestimable value and stimulus to students, and made an everlasting mark in American Ornithology. It is indeed hard to imagine what the science would be like in this country—and what the state of our bird world—had he not lived and wrought, and become a demigod to the ardent youth of the land.

I had not meant to write an essay on Audubon—and shall not continue! I do want to thank you for the truly great encouragement of your letter, and to assure you that if there is any way I can help you to develop your bird-painting, I am anxious to do so.

Faithfully and gratefully yours,

Louis Agassiz Fuertes

This first letter from Fuertes I read with understandable excitement and elation. That he had replied at all surprised my parents a little, but it did not surprise me. I had felt for a long time that Fuertes would understand me; that he might, indeed, be the only other person in the world who felt as I did about birds. He would know all about my desire to draw them well, to capture their beauty. I read the letter over and over. How strange it was to be addressed as "Mr. Sutton"!

What surprised me most about the letter was the forthright, the almost defiant, defense of Audubon. Teen-age upstart that I had been, I had tossed off derogatory statements about Audubon, statements aimed primarily at proving how superior Fuertes was to his famous predecessor. I had said my callow say without ever having owned, or even carefully studied, a volume of good Audubon reproductions, without ever having seen a single hand-colored lithograph from one of the magnificent Audubon folios. When I thought over what Fuertes had to say, I knew that he was scolding me a little, putting me in my place.

When I contemplate today what Fuertes wrote, I am glad that I went overboard as I did in my first letter to him, for my pronouncements, half-baked though they doubtless were, called forth a wonderfully articulate brief on Audubon—a brief that but for my letter might never have been written, a brief that was the more meaningful for having come from one who, perhaps more than any other person living at the time, was in a position to know what the problems of bird portraiture were. How I thrill as I read today what the great Fuertes had to say about Audubon!

What Fuertes meant by the "stilted, tight and unimaginative old David" may need explanation here. According to some of his many biographers, Audubon studied for a time, when very young, with Jacques

Louis David, the court painter to Louis XVI and Napoleon I and the founder of what has been called the French classical school.

In my first letter to Fuertes I must have said in so many words that I wanted to know him personally, to study with him if possible. I had not thought, really thought, about what traveling from Bethany, West Virginia to Ithaca, New York might entail either financially or otherwise. Nor had I thought at all about how further interruption might inconvenience Fuertes. I had simply said my youthful say, overflowed as it were, and Fuertes had responded—generously, almost affectionately, as if he really wanted to know and help me.

How I wondered whether Fuertes had seen that Roadrunner drawing of mine in *Bird-Lore! He must have see it,* I told myself. *How could he keep from seeing it, when some of his own work was appearing in the very same issue of the magazine?* I gave the little picture another long, hard look. How crude, crude, crude it was! The nearer wing was a wretched blur. The bill did not open correctly. Especially offensive was the small rock that I had added to hide the poorly delineated right foot. To be sure, there were ten tail feathers, the right number. Yes, the eye's pupil was, as it should have been, encircled with a light ring. Yes, the plumage of the back was rumpled, just as it had been in the startled bird. The picture was not all bad. Yet Fuertes had said not a word about it. I was puzzled, not hurt exactly, but puzzled.

I had a few drawings, made the preceding fall, that

I felt were good enough to send to Fuertes for criticism. All of these were of small birds that I had shot with .22 caliber shot-shells. Two of the drawings were of special interest to me, for I had made them partly to fasten in my mind the differences between Bay-breasted and Blackpoll warblers in their first winter plumage. Both species were new to me and both had been extremely common the preceding September and October. They were not easy to tell apart in the field or even, for that matter, in the hand. Among the other drawings, all in watercolor and all life-size, were one of a Magnolia Warbler in fall feather and one of a Red-winged Blackbird.

In his second letter, Fuertes discussed some of my many shortcomings, but not one word was aimed at discouraging me:

LOUIS AGASSIZ FUERTES
CORNELL HEIGHTS, ITHACA, N.Y.

Mar. 30, '15

Dear Mr. Sutton—

I was somewhat chagrined when I received your letter of the 15th, to realize that I had not acknowledged in any way the receipt of your drawings. They came in due time and good condition, and I was waiting, before writing you fully, until I could find an hour of daylight in which to study and criticize them in such a way as they deserve, and as would be helpful to you. I have been greatly rushed this winter, and just at the present time I am in the middle of a long series—some 72 little plates—of game birds, for a later number of the Geographic Magazine, for which the publishers are clamoring. Hence my present rush.

I very much appreciate your warm praise of my pictures. Of course I take them very seriously, and it would be insincere of me, under the circumstances, not to warm up under expressed appreciation of them. They are all far below the little pictures I have in my mind, and if they express any of the beauty or charm or feeling that I have in mind for them it is only because the imaginary picture is so surpassingly lovely that part of its charm penetrates through the heavy and stubborn medium of my clumsy hands, in spite of them. My only salvation is that nobody can see what I had in mind for my pictures! If we could paint without restraint or manual obstruction the beautiful pictures we plan, how we could make the world open its eyes!

To come at once to the criticism: First of all your work shows great promise, and you should by all means stick to it and work for the highest ideal you can summon.

Technically, I think you can work to advantage with a somewhat drier brush. It is impossible to exactly control either color or outline, unless your brush is sufficiently dry to avoid leaving a drop at the end of a stroke, & to distribute color evenly through the whole stroke. Take a bird with a white belly and study, locally, the color of its shadow on the white. Put colored things so they can reflect in this shadow, and see what a mirror of adjacent colors white becomes in shadow. Your shadows lack in both color and in depth. It would be good exercise to paint, literally, a bird-skin or a dead bird with the one aim of getting every point on it in exactly the proper relation in both color and value (value equals the relative amount of lightness & darkness, irrespective of color) to every other point. This will give your object—or bird in this case—solidity—and rotundity—, and is the only thing that will. Every object or landscape in the world—in fact anything you look at or isolate for a picture—has one deepest dark spot and one highest light, and every other spot has its exact relation between these and to every other spot. This is the main thing to know, in painting, and to properly recognize and master this analysis is the principal result gained by an art education. You are supposed to start with the appreciation and beauty-sense, and these things develop automatically as education progresses and experience accrues.

To take an example, let us analyze the fall Magnolia Warbler picture. Your drawing is nice, the conception pretty, graceful and very characteristic. The coloring is, locally, correct. But it so happens that local color (by which I mean the exact tone & color of a given part of the bird, seen in most favorable and analytical light) only tells at its face value and color rarely. On top, where much light hits it and sprays off, it is greatly lightened and "grayed"; underneath, where an inevitable shadow falls, it is greatly darkened, and also reflects color and life from surrounding objects. If you realize at once what it has taken students all these centuries to wake up to—namely that all animals are dark on top and pale below for the sole purpose of overcoming or compensating this inevitable top-light and bottom shadow, your main difficulty will be solved. It is possible—and frequent—that white tells darker than black; that is, if white on the underside falls into sharp shadow it is apt to be actually darker in value than black on the top, shining in or bathed in much light. These things must all be studied conscientiously and with a wide-open mind, as all our natural ideas are apt to be violated when actual conditions are studiously examined. The way to do is to take any object you want—preferably in simple design & colors at first, and do a picture of it three or four times, in different lights and positions, to accustom yourself to its changes, and get away from your traditional prejudices that a given surface is always to be represented by the same pigment or in the same manner. In other words, loosen up, be unafraid to try and represent what you see. It won't hurt anything if it does turn out awful: it may turn out right, and in any case you've learned something, either to do or not to do. And that is progress, while plodding in the track of prejudice without courage to kick out & see what happens is not going to develop anything better than it has already developed.

To point what I have tried to say, I have taken your own little Magnolia [Warbler] drawing, and made a transcription of it which I am enclosing, with your other pictures, and returning to you in this same mail. It is not perfect—nor exactly intended to be—but it is designed to show you a little how white can look whiter when it's made of black and Prussian blue than when it is left blank (when it looks like a hole) and how black in shadow looks just like black in light not when the two are just alike, but very different in order to take their proper values in surfaces opposing the light differently. Except for the color, the black on the upper wing-coverts equals the white under the tail. Compare the relatively adjacent white and black.

These suggestions should prove helpful to you. Work all for study, and none for either exhibition or commendation, at least for a while. As long as you fear criticism you will hedge at making experiments and downing tradition. So don't work for anything but your own progress. Send me your results and I will help all I can. Send your mistakes preferably to your successes. If I can help you it won't be by praising you but by showing you how to overcome your difficulties. To do that I must see what they are. Learn not to fear contrasts in the same field of color. They are hard, but they occur, & are consequently to be reckoned with.

I would like to know something about yourself. How old are you? What are your surroundings & circumstances? Do you wish, some day, to make this work your livelihood? And if so, have you any other source of income in case of tight times? These are all

Fuertes drawing of a Magnolia Warbler sent to Sutton with the March 30, 1915, letter from Fuertes (which see) but not signed for several years. The inscription was added at Sutton's request.

pertinent questions, tho' they may seem to you impertinent. I hardly think I need tell you that no other spirit than that of professional sympathy enters into my asking them. I realize, probably better than anyone else, because it is my sole profession, what its demands and emoluments are, and its risks as a livelihood. As an avocation I should put it beyond anything I know of in value: as a vocation I should certainly think long and hard, and prepare for many sacrifices in entering the lists. I have been able to command about the same annual income as I should have done as a college professor with the same grounding. I had the advantage of being early in the field, and having an excellent teacher, in Abbott Thayer, to help me to overcome my early difficulties, and an excellent executive friend in Elliott Coues, who badly needed an illustrator just as I got ripe. I have, of course, no university connection to insure me a Carnegie pension when I begin to slide down the far side. I have been lucky beyond my deserts, and I have also worked hard and long and still am. That is no hardship in a profession of such boundless opportunities for the hand and the imagination to work together. But it is not exempt from the temptation to "pot-boil," and unless you are resolved to be forever your own most relentless critic you won't be able to hold your banner high. I don't know, yet, whether you've got the thing in you so hard that it will prove your dominant passion right through your life or not. If not, I mustn't encourage you to give your life to it: if you have, I wouldn't be able to dissuade you anyway. I ought to tell you, probably, that I was considered mildly insane, although myself the only sufferer from the results of the malady.

I had not intended to write you a book on this matter, but I want only to impress you with my conviction that the field, as a sole profession, is a risk. As a pastime that may later develop into a valuable asset, provided you prove to have the persistence to develop the requisite skill, and above all, to have that insight into nature and that sympathy and understanding of birds which will make your pictures real contributions that the world must have, you may safely launch into the work and fear no rivals and no depression of your business. You'll never be rich, and you'll surely never be poor.

Very sincerely yours,

Louis Agassiz Fuertes

Here is a sketch to show some of the principles I have tried to describe. Notice that

1——the darkest tone (except holes) comes not on the edge away from the light, but between the lighted side & the shadowed side.

2——the reflected light on the shadowed side not only rises in value, but alters in color, due to the surface that reflects this light.

3——local color is almost absent, & the whole thing

is largely reflected or induced color from surrounding objects—

4——*High lights and deep darks: (squint for them if they are not apparent in the mass of details seen distinctly) There is always one of each, between which every other part must take its proper place in the scale. Here the high-light is the left eyebrow, & the nose holes form the deep dark. Excepting holes, everything in the lighted area is higher than anything in the shadowed area, & vice versa. This skull, by the way, is the artificially distorted skull of an old Flathead Indian, hence its flat forehead & stampeding brain case. In profile thus——*

Take some unfamiliar object like this and make similar studies, as pure training of your eye. We are naturally almost blind to all but the most blatant & contrasting values, & even great painters—especially of the earlier schools & periods never did get discriminating in this essential matter. Study, if you can get at them, the works of Geo. Deforest Brush & Abbott H. Thayer, in this connection. Many old school painters understood contrast thoroughly, & played it like the masters they were. Few if any understood—or correctly rated—the importance of a perfect balance of values, & I think the reason many greatly esteemed

deep darks: (sequence for them
rent in the mass of details
There is always one of each,
ery other part must take
e in the scale. Here the
eft Eyebrow + the nose holes form
Excepting holes, everything in
s higher than anything in the
ea, + vice versa.
ag, is the artificially dis-
n old flathead Indian,
forehead + stampeding
profile thus-

n object
imilar
ing of
terally
but the most blatant + con-
rn great painters - especially of the

Page from letter written by Fuertes on March 30, 1915.

works survive is because critics and amateurs alike are so untrained in the detection of this cardinal principle. Color subtlety is simplicity itself compared to value subtlety, which includes every possible element of color variation as thoroughly as does white light itself. Great pictures are often full of ingenious tricks of color to overcome the failings of the painter to achieve his values.

Your redwing study is a valuable case in point, & a difficult one. Do another, with especial regard for this matter of values. Don't be afraid of failing & do it over several times. That is the best possible study, & should not prove a bore if the 2nd is better than the first, etc. Study it all out, after you've made your drawing, & before you start to paint; paint your high light & your deep dark, & adjust all the rest properly between, & then go to it.

L.A.F.

This second letter—a long one—was almost too much for me. I read it through several times at one sitting, then returned to certain paragraphs and sentences to be sure that I was understanding what I read. Never before had I been exposed to such phrases as "high lights" and "deep darks." Never had I even thought of the colors of shadows or of "local color" as opposed to what that very same color became when in shadow or when influenced by the colors of objects close by. The idea that white-in-shadow was a color rather than a mere lessening of whiteness, or that black was never really black when struck by sunlight, had never crossed my mind. Even as the "primrose by a river's brim" had never been anything but a primrose to the poet's vagabond, so white had always been white to me, and black, black. The two had been absolutes, though of course I had known that there were such variants or "shades" as satiny white, ivory white, blue-black, and velvety black. I had never really thought about values, as such. Each time I read the letter I found myself agog with ideas, some clear, some not so clear, but all of them exciting, even disturbing.

The letter said, in so many words, that my work showed "great promise." That was important, all-important. The letter urged me to keep trying. It made some sound suggestions, too, among them that I draw a dead bird, not a living one, by way of seeing for myself how light raised "values" as it fell on certain parts of the plumage, producing these "highest lights" where it was strongest and, by the same token, the "deepest darks" where it was weakest or not present at all. Putting down a record of these "highest lights" and "deepest darks," with all of the intermediate shades correctly worked out and positioned, would result in a three-dimensional quality that my drawings did not have.

I laid a dead Blue Jay on its back with head toward me on the table and got to work. As I finished the pencil outline I realized that I had never before dealt so directly with foreshortening. In the past I had foreshortened wings, tails, and other long parts of birds not with specimens before me but from memory of pictures that I had seen and considered good. I had, in other words, been making pictures from pictures, not from the birds themselves. When, in the past, my drawing of certain parts had not "measured up," I had used the eraser and drawn again, and I had continued with eraser and pencil until I had made the parts acceptable—i.e., in agreement with pictures I had seen. This time, *this time,* I had drawn with a model in a strange position before me, not a photograph, not someone's drawing, but the bird itself. The experience had been profoundly revealing.

When I added color to my pencil outline I perceived at once the truth of what Fuertes had written. What I could see of the pure white bar on the jay's wing was now in shadow, deep shadow. What I saw of the bar was distinct, partly, I suppose, because I knew it was there. I knew the bar was white; yet what I could see of it, shadowed as it was, was far less white than that of the middle of the belly, a part of the jay that was receiving much light. That white wingbar was, indeed, almost as dark as the shadow on the table under it. Furthermore, it had almost the same *color* as that of the shadow on the table. It was no longer truly white at all; it was, indeed, almost anything but white! Heaven help me, was I seeing things? Yes, I was; quite literally I was, and for the first time in my life.

A point to be made here is that Fuertes had said not a word in his second letter about my failure to foreshorten properly. He had been concerned with the flatness of my work, with its failure to be three-dimensional. So what I discovered about foreshortening as I drew the dead jay was accidental—a kind of by-product. And it called attention to an important fact, namely, that in every drawing I had thus far made I had—perhaps quite unintentionally—avoided the difficult. Never had I drawn a bird with its tail or spread wing pointed toward me. I had been content with "easy" attitudes, profiles. The more I thought about this aspect of bird portraiture the more I wondered whether any bird artist could ever make the bill of a Great Blue Heron *look* long if he drew the bird with bill pointed straight at him!

The drawing that Fuertes had made for me of the "artificially distorted skull of the old Flathead Indian"—the item discussed in the letter's all-pencil addendum and sent along with drawings being returned to me—was a small watercolor done in great detail in brown tones. I found it hard to believe that Fuertes had made this, for it was not of a bird or even of part of a bird, yet obviously it was extremely good—and beautiful. I enjoyed looking at it, finding that it illustrated, almost dramatically, how light seemed to fall on objects, piling up like snow where parts protruded and missing altogether the cracks and holes.

The "transcription" that Fuertes had made of my Magnolia Warbler drawing thrilled me as nothing in my whole life had ever thrilled me. What impressed me most about the drawing was the way in which it proclaimed know-how. Nowhere could I find in it, even with the closest inspection, one trace of rubbing-out or fussing-over. The delicate tarsi, the eight little toes (not drawn separately by any means, but all there), the bill, the eyes, all had been put in place by someone who knew precisely what he was doing. I had the odd, odd feeling, as I studied the sketch, that the warbler itself—not the drawing, but the bird itself—would have a hard time matching the beauty and verity of the image on the paper before me.

Some of the letter's clearly stated practical suggestions I never got round to following. Never did I try drawing "three or four times" an object of "simple design & colors . . . in different lights and positions." Never did I re-do my study of the Red-winged Blackbird. Nor did I consult the works of George Deforest Brush and Abbott H. Thayer—as I should have. I was living in a very small town. It never occurred to me that the great Carnegie Library in Pittsburgh, only sixty miles away, might have in it the very books Fuertes wanted me to see. Besides, we had no car, so Pittsburgh might as well have been on another planet insofar as I was concerned.

Fuertes had become more than casually interested in me and my problems. This I could tell from the tone of his questioning. I set to work as I had never worked before, determined to show this wonderful man that he was not wasting his time on me.

The third letter from Fuertes, addressed once more to "Mr. Sutton," dealt largely with shadow-color:

LOUIS AGASSIZ FUERTES
CORNELL HEIGHTS, ITHACA, N.Y.

June 5, '15

My dear Mr. Sutton—

I have delayed writing you for the same old reasons—I've been very busy & haven't felt I could write you as fully as your hard work and gratifying results warrant. And even now I won't be able to go far, but I want to tell you that I received your pictures and to tell you that you are going ahead nicely. Do persevere and work more from fresh material: the Song Sparrow is one of your very best.

The trouble with your shadowed Chestnut side [*Chestnut-sided Warbler*] *is only that you haven't quite arrived—you are headed quite right. Shadow-color is not easy at all to analyze. You have to un-know everything you have always thought & start with a clean slate—or an absolutely open mind. Analyze with a perfectly unbiased mind. A good way to help is to take a piece of white paper, cut a small round hole in it and study your color locally by isolating it from the rest, & comparing it with the white paper around the hole. Do this outdoors, too, & see what surprises you get as to actual color of hills etc. in the distance. Or tear a small hole in a big leaf & look at the skyline through it. All these things will help to clear your mind of preconceived notions of color—especially it will show you that local color—(by which I mean the exact color, analytically, of any given part of your whole subject, isolated from the rest) is practically never unmodified by either warm light or cool shadow; in the latter case, too, it almost invariably absorbs color either reflected from nearby objects or blue absorbed from the atmosphere. You don't have to have blue sky in the latter case. That is why your warbler looks muddy. For exercise, try greatly exaggerating the colors thus reflected, as well as the lightness of the parts that catch light. I don't think it will be dangerous for you to do this, for your eye and hand are both sufficiently literal to hold you safe. But I doubt if you can grasp the extent of this effect of atmosphere—(to lump the whole gamut of complications) unless you take a good swing out into it. Put your bird where the sun hits it, & paint it—exaggerate if you can—the full relation of the light side & the shadow side first, and, secondarily, work in the surface color and markings. Take a fair sized subject, so you can study it freely: a stuffed hawk is good. Don't be afraid to spread paint.*

Your two Chestnut-side studies are just the kind of thing that will help you: the same theme under different light conditions. If you have difficulty in adjusting your markings to your broader surfaces, half-shut your eyes, and thus exclude the complicating details—both of pattern & of values. Divide your bird along its "twilight" line—thus—[drawing of small bird] and keep night from day.

Usually you will find the darkest line just beyond the line—not at the edge, for some light always comes adjusting your markings to your broader surfaces, half-shut your eyes, and thus exclude the complicating details — both of pattern & of values.

Divide your bird along its "twilight" line — thus — and keep night from day.

Usually you will find the darkest line just beyond the line — *not* at the edge, for *some* light always comes around the other way, (usually as a reflection from something else.

In sunlight, *everything* in the lighted side — black included, will prove lighter than *anything* in the shaded side, *white* included.

These are all basic principles that I may have pointed before. I can't quite remember just what I have written before.

You are doing nicely and must not be discouraged. At your age I did far less direct and sprightly things. Send me some more, and if you come to Ithaca let me know, and we'll talk it all over with you. Yours, L. A. Fue

Page from letter written by Fuertes on June 5, 1915.

around the other way, usually as a reflection from something else.

In sunlight, everything in the lighted side—black included, will prove lighter than anything in the shaded side, white included.

These are all basic principles that I may have posited before. I can't quite remember just what I have written before.

You are doing nicely, and must not be discouraged. At your age I did far less direct and sprightly things. Send me some more, and if you come to Ithaca let me know, and I will talk it all over with you. Yours,

L. A. Fuertes

Had Fuertes not said to me in this, his third letter, that I was "going ahead nicely," might I have decided to give up drawing birds? Questions of this sort cannot, of course, be answered. Insofar as I myself was concerned—irrespective, that is, of my parents' feelings—the fact that Fuertes was criticizing my work was proof enough that I was getting somewhere. For further assurance all I needed was another glance at the Magnolia Warbler sketch Fuertes had made for me. How I idolized that little picture!

During the spring of 1915 I had sent Fuertes a batch of drawings, three of which won comment from him—one of a Song Sparrow that he considered among my "very best," and two of a Chestnut-sided Warbler, one of which he called "muddy." I cannot remember whether I had done the Chestnut-sided Warbler studies from the same specimen; nor can I remember whether I had made a point of doing them, as Fuertes obviously believed I had, "under different light conditions." In any event, that word *muddy* bothered me sorely. *Muddy* was a strong word, one connoting something disagreeable, something ugly. If I had perpetrated a picture that deserved the word *muddy*, then drastic moves from me were in order.

I studied carefully the sketch whose muddiness Fuertes had brought into focus for me. At first I felt that the picture's shortcomings, whatever they might be, had been over-stated. Nothing about the bird as drawn looked as if it had mud on its feathers. Was *muddy* the right word?

I punched a hole about the diameter of a lead pencil in a piece of paper and, following the explicit directions given me by Fuertes, I squinted through the hole first at the drawing, then at my skin of a Chestnut-sided Warbler, directing my attention to the parts that were heavily shadowed, the lower belly and under tail coverts. At once I realized what Fuertes had meant: insofar as those parts of my subject were concerned, the beauty had been lost in the drawing. The illusion of featheriness was missing. There was no smoothness, no softness, no hint of transluscence. I knew that I had put down one color and decided that it was wrong. Then I had put down another color, hoping to correct the first. That had not helped. Then

I had decided to scrub it all into acceptability, with a result that looked—well, alas, it looked as if soot had been mixed with the paint. On concluding my investigation, I decided that *muddy* was not a strong enough word. *Gravelly* might be better. How was one to avoid any such *gravelliness* as this?

The Sutton family was living in a small, one-story house at this time, and I had no room of my own. Professor Milton Hover, who taught Biology at the college, let me do my drawing in the Biology Laboratory on the second floor of Oglebay Hall. To help with the family's finances, I mowed lawns and lathed a house that was being built along the edge of town.

That trick of looking at things through a small hole in a piece of paper intrigued me no end. What surprised me most of all about the trick was that I could use it on pictures. If the picture was good, its various parts fell precisely into place: what was near looked near, and what was far, far. I could use the trick in judging my own drawings. If what I had drawn was correct in perspective, then parts that represented what was far off looked that way. More convenient than a hole in paper was a peep-hole I could make with either hand. This I used as I studied pictures wherever I happened upon them—a huge oil of the Holy Land in the college chapel, a calendar on the wall in the post office, pictures in houses here and there in town. My habit of squinting at pictures through the peep-hole in my hand gave people the idea that I considered myself quite an art critic.

LOUIS AGASSIZ FUERTES
CORNELL HEIGHTS, ITHACA, N.Y.

Sheldrake, N.Y.
Aug 3, 1915

Dear Mr. Sutton—

Your letter came a week or more ago, and with it the little parcel of drawings. I haven't at present anything more to tell you, but repeat my urging to go on studying, by the means I outlined, to arrive at a keener analysis of shadow colors.

Quite the best in this lot was the buzzard skull. You got a good deal out of that study, and it is good for you to study other things than birds once in a while. Even here, I feel sure, you were either unable to see and analyze the color or a little afraid to give it full rein. It is well to remember that most if not all water colors dry duller & lighter—more like a "bloom," than the color they go on when wet, so you may safely overstep a little—(or a good deal, in study, to try your wings.)

I meant you to keep the sketches I did for you. They are to point certain principles, & if you study them in that light will do you more good than anyone else.

Continue to send me your studies, and I will criticize them gladly. In this lot the chief criticism in all is rather the <u>*color*</u> *of your shadow than its distribution. Keep at it; experiment, and don't be discouraged at failures.*

They are all constructive: it's as important to learn what <u>not</u> to do as what <u>to</u> do.

Very truly yours,

Louis A. Fuertes

In the summer of 1915 I had sent Fuertes the "little parcel of drawings" mentioned in this, his fourth letter to me. Most of the drawings were of small birds that I had shot along the Buffalo or along one of that pretty, unspoiled stream's major tributaries, Castleman's Run.

Among the drawings was a detailed one of a specimen brought with me from Texas, the skull of a Turkey Vulture. This study of something "other . . . than birds" Fuertes liked. He was, I now know, more than justified in saying that I had been "unable to see and analyze" this skull's shadow-colors. The fact is that I had not even tried to see these colors. The shape of the bones and the way they fitted together was what had interested me. Shadows helped to make clear which parts were closest and which farthest, that was obvious; and as I recall the experience of drawing the bones I made not the slightest attempt to figure out what the colors of the shadows were. The position of those shadows and whether they were light or dark was all that mattered.

Ever since receiving from Fuertes the Magnolia

Warbler sketch and the sketch of the old Flathead Indian's skull I had felt that the two pictures should some day be returned to the artist. A statement in this fourth letter made clear that I was to keep the two pictures. It also encouraged me to send more of my work for criticism. "Continue to send me your studies, and I will criticize them gladly" were the exact words. "Keep at it; experiment, and don't be discouraged at failures. They are all constructive: it's as important to learn what *not* to do as what *to* do."

How applicable to all of life were exhortations such as these! Those elusive shadow-colors: I would try harder than ever to see them, analyze them, and get them right!

LOUIS AGASSIZ FUERTES
CORNELL HEIGHTS, ITHACA, N.Y.

Nov. 12, 1915

Mr. George Micksch [*sic*] *Sutton*
Bethany, W. Va.

My dear Mr. Sutton—

I have so long put off answering your letter, and acknowledging the drawings you sent, that I am almost ashamed to write now. And, as it is, I can only write a short note to prove that I'm neither dead nor sleeping, but only exceedingly busy with contracted work, and so unable to do more than let you know that I did receive, in good shape, all the studies and the very appreciative letter you sent in September.

Your work, while I cannot criticize it in detail here, shows great and logical improvement. I can only say, in general, that your work is bearing out my early conviction, born of your drawings and of your point of view, that you could go far with it, and make it a source of much pleasure to yourself and to others. As opportunity presents itself, make more and ever more drawings, making each one more truthful and studied than the last.

I will, as soon as this burden of responsibility gets off my shoulders, give your last lot specific criticism & return them to you. In the meantime, go on with your work. You have fallen into no bad habits in your drawing, but have still much to learn about local and reflected color, and their combinations. Don't get discouraged, and draw, draw, draw.

Very truly yours,

Louis A. Fuertes

In September of 1915 I entered Bethany College, not as a full-fledged freshman, for I had not quite finished High School, but as a special student enrolled in certain "prep school" courses. I was employed as a janitor on the campus, my chief responsibility being

Oglebay Hall, the "Science Building" that had recently been added at the end of the long, stately, wonderfully handsome "Old Main." On the slope north of the campus and back of our house was an apple orchard owned by the college. A trail through this and the woods beyond led to a shallow cave known amongst the youngsters of the village as The Amphitheatre.

My schedule was full, what with classes and the mopping of floors, but I managed to get into the field almost every day and some of the birds that I shot for my growing collection I also drew. I did all of my drawing in the Biology Laboratory in Oglebay Hall. There the light was good and at certain hours there were no interruptions. At the height of bird migration I noted, even as I had the preceding fall, how abundant certain warblers were, especially the Bay-breasted, Blackpoll, and Tennessee. The Cerulean Warbler, a species that summered in considerable numbers in the mature oaks on the hilltop above campus and town, I never heard singing in the fall, and only infrequently did I see it at that season.

The fifth letter from Fuertes acknowledged receipt of several drawings, at least two of which (immature Rose-breasted Grosbeak, immature Blackburnian Warbler) I had done from unskinned specimens. One of an adult male Cardinal, however, had been done from a bird that I had carefully mounted, and this study was much more of a picture than any I had thus far sent for criticism. It had a scenic background that showed, above creek and hillside, a sky filled with white clouds.

Fuertes was exceedingly busy in the fall of 1915. I believe he was at work on a series of drawings for *The National Geographic Magazine*. It was kind indeed of him to write me as he did on November 12, "You have fallen into no bad habits in your drawing, but have still much to learn about local and reflected color, and their combinations. Don't get discouraged, and draw, draw, *draw*."

Much to learn. Shadow color. Local color. Combinations of local color and reflected color. Was I really moving ahead as a bird artist?

LOUIS AGASSIZ FUERTES
CORNELL HEIGHTS, ITHACA, N.Y.

Jan 9, 1916

My dear George—

I see by your letters and your pictures that you will not mind if I give you your more personal name, and it puts us on a much better relation to work with.

First of all, I am going to ask you to thank your father for his kind message at Christmas time. I was very much touched to receive it, and by his inscription. I have not as yet read it, for I lent it at once to a sick friend who would, I knew, get much enjoyment and encouragement from it, and it has not been returned yet.

Your two warblers are lovely little drawings and very delicately done. I think the criticism you give the B. & W. [Black-and-white Warbler] as to her eye, is unfounded. I find the proportion quite perfect. I should, however, criticize it as being a little unfelt as to the values of the blacks on the top, and perhaps, too, as to the shaded underside. But you are coming on, strong, with your drawing and your color too.

I feel pretty sure that the blacks, which are quite glossy, would reflect a lot of blue lightness off the back, and result in a unity of surface in place of the sharp ridges and troughs that result from your juxtaposed real black & real white. You may remember, in life, the distinctly blue look of a Black & White Warbler: it is due to partly sky-shine & partly the blue-black, (locally) of the feathers themselves. I should recommend your using lamp-black instead of ivory or illustrator's black: it is more dead-black, and washes into a cool stone-gray instead of a warm brown-gray.

Your little drawing is exquisite and dainty, and your technical handling of feathers is a vast improvement over anything I have seen from you. You will find (as a suggestion only) that to do the fine white borders to primary feathers etc. a very direct and satisfactory way is to ignore them altogether while putting in your black ground-color, and later, when it is fully dry, putting them on with a pen & Chinese white. I do not advise "body-color" for general use: when you have small local contrasts, however, it is sometimes the best way to your result.

Take all the time you need, by all means. But

also practice handling larger areas freely, with a brush-full of color, & learn how your different pigments act with free handling. You can do better with a fairly large brush, well pointed, than with a little one, even in fine places, for it holds and delivers the color much more evenly. They are more expensive, but pay.

Your ♀ Cape May [Warbler] is a beauty, too. In some points it is superior to the B. & W. [Black-and-white Warbler]; here, however, I have a feeling that the restraint is too great. I think (but look it up for yourself) that the olive is a bit too light and clear, and the gray around the head & face correspondingly so. This results in denying the yellow (about right) its proper value from contrast, & making the bird look a little translucent, instead of solid, but lighted on the surface. Think of light as a white powder, sprinkled on the top surfaces of your birds. For that is almost exactly the effect it has on local color. But no two female Cape-Mays are just alike, and my criticisms may be ill-advised, as I haven't of course seen your specimen. Mine are both (I have 2) much browner-olive above, but both are spring birds, in somewhat worn & faded plumage. Your drawing is very delicate, and your values, while a trifle light, are beautifully preserved and balanced. Where was this bird shot?

This I have not touched. The B. & W. (which I return today with all your others,) I have just touched, to show you how very near you came to getting it right! I have let up the black on the crown & the parallel plane of the back and pointed the accents in the breast & side streaks—otherwise nothing. You can see at once how it jumps into reality. I have compared it with a ♀ in my collection, and find my <u>*color*</u> *criticism unjustified, for the male is much brighter and bluer black, and tinged my impression of the species.*

I would much rather have one or two studies, like these, on which you have taken your time, than a lot less carefully studied. You are fast coming to the point (but you haven't quite arrived yet) where you can paint what you see. So keep studying with your eyes, analyzing color wherever you see it. You struck a hard one in the Cardinal. Red in shadow is very complex, being intense on its own account & still reflecting, strangely, opposing colors. Fight it out some time when you feel good and strong: it still makes me guess and fight.

It will surely encourage you to compare the last two sent with the others I am returning.

As a general criticism, I think you show the wing too far up. The shoulder or scapular feathers usually if not always cover the base of the wing, at least over the inner few coverts, & most of the shoulders, in life. I have the feeling that all these birds have the feathers brushed away to show the whole wing.

Also, your shadows are still a little dull and muddy, tho' greatly improved. Compare the young Blackburnian W. [Warbler] dated Aug. 4, on this score. A very nice drawing, with the shaded area well studied as to concentration, reflected border-light etc. but dull and colorless, thereby becoming in appearance not shadow, but soil. The young Rose-breast [Grosbeak] shows the same trouble. Also the tarsus is considerably too long. They are shortlegged birds. My specimens all have a lot of rich buff and umber through their general plumage; was your bird as pale and dull as the picture?

I am very glad indeed to have your photograph. I have naturally wondered a good deal what kind of a looking boy you were, and am glad to know. I hope you may find it possible to come to Ithaca some time: perhaps next summer session. If you do, come and look me up as soon as you get settled. Perhaps I could help you with that part of it, if you come up. Let me know, anyway.

I am enclosing a little print of one of my hawk pictures, and also a rather poor picture of myself.

You are certainly improving very satisfactorily, and should feel much encouraged, for your improvement is the important kind that comes with real study. Send me more at any time you may have it. I would concentrate, if I were you, on the matter of making very shaded white areas transparent and colorful. The color, by the way, is usually cool or blue, and the reflected light along the edge is usually warmer. Be careful your warm area is lighter than the cool part, and not muddy. It may take some experimenting to land on the right colors—doubtless will.

Very sincerely yours,

Louis Agassiz Fuertes

This, the sixth letter from Fuertes, was the first in which he had addressed me not as "Mr. Sutton," but as "George." I really wanted Fuertes to know me as George. I didn't feel quite comfortable when he called me Mr. Sutton.

The letter was prompted by receipt from me of two small drawings that he obviously liked. One of these, of a female Black-and-white Warbler, I had made during the preceding fall, while migration was on. The other, of a Cape May Warbler, I had done on December 7, when the bird had been brought to me, in a weakened condition and minus most of its tail, by a neighbor.

The criticisms Fuertes offered on these two studies were extremely helpful, partly because I understood them so fully. I had noticed the glossiness of the black parts of a Black-and-white Warbler's plumage, the glossiness as well as the bluish tinge, features that were more evident in a fully adult male specimen than in an immature male or in a female of any age. I was sorry that there was no way for me to compare directly the study of the bird as I had made it with the study as Fuertes had "just touched" it with his magic brush. Needless to say, I scrutinized the picture as it was when returned to me, trying to decide precisely where I had made the black parts of the crown and back too black and the breast and side streaks either too black or not black enough. I studied the touched-up picture a long while. I couldn't help feeling that it was really good. It had feathery softness. Light seemed to be falling on it. It was not quite *my* picture, of course, but I had had a good deal to do with making it.

As for the "general criticism" of my tendency to place a bird's wing "too far up" and to "brush away" the scapular feathers so as to show too much of it, I was greatly perturbed. Fuertes was speaking now not of the two warbler drawings, but of a batch of sketches I had sent the preceding late summer or early fall, drawings mentioned, but not criticized at all, in his fifth letter (November 12, 1915).

Here were errors that I would have to correct not through study of prepared skins or freshly shot specimens but through watching live birds more closely than I ever had. It was winter at the moment. The Tufted Titmice, Carolina Chickadees, and Cardinals that I saw every day were properly fluffed up from the cold. Their wings were not, I had to admit, very conspicuous much of the time. The wings of a Carolina Wren that I followed about for half an hour were so covered by breast and belly feathers from below and by shoulder, back, and rump feathers from above that I could hardly see them at all. Perhaps, in my determination to record such diagnostic features as wing-bars, I had been overdoing them.

The thought suddenly hit me that I had more than once been impressed with the fact that a swimming duck's wings were so covered by shoulder, side, and flank plumage that their really showy feature, the speculum, was completely hidden. I had first noticed this in Fort Worth, Texas, at the zoo in Forest Park, where the shiny green patch on the wing of a drake

Green-winged Teal had, to my surprise and disappointment, never been really visible unless the bird stretched or preened its wings.

And my Rose-breasted Grosbeak's leg was "considerably too long"! How could I have made that bad mistake? The answer this time was simple enough: I had been careless. I had drawn from imagination or memory rather than from any sort of measurement of what was before me. From now on I would, if necessary, hold my model's leg close to the paper beside the drawing to make sure that I was getting its length right.

I never tried using a pen and Chinese white in making "fine white borders to primary feathers." I had no white of that sort at the time and didn't know where to obtain it. Furthermore, I had no pen that I felt would be capable of doing such work.

The "little print" that Fuertes sent me was a tinted photograph of his drawing of an immature Goshawk. I have never seen the original and have no idea for whom it was made. It was a wonderfully vivid portrait. The eyes had fire in them and the feet, power.

The photograph of himself that Fuertes sent showed him standing near a drawer full of bird specimens.

LOUIS AGASSIZ FUERTES
CORNELL HEIGHTS, ITHACA, N.Y.

Jan 21, 1916

My dear George—

I cannot now write you much of a letter, but I want to acknowledge your very good letter of the 14th and to reassure you a little in case you find it possible to come to Ithaca in the summer. My own plans are not at all formulated as yet, and I cannot therefore quite certainly assure you that I myself shall be here, tho' I expect now to occupy a summer cottage on the lake only about 18 miles from Ithaca, and easily reached. In case everything goes as you hope, and you come and I'm here, the matter of "lessons" need not worry you at all. I have no secret processes or other patents on this profession, and I should consider it a privilege to pass on to you what little of the help and encouragement I could of all that has been so freely lavished on me by my good friends in days and years now past, and if I should myself seek reward for so poorly doing what was so richly done for me, I should in truth feel like a worse parasite than I hope I shall ever have to feel! If you should come, you will be allowed to pay for everything you get, except such help and criticism as I can give you while you are here: <u>that</u>, at least, you cannot pay for, because it will be no trouble at all, and I shall be, as I have said, only too happy to help you along. So don't embarrass either of us by over valuing it, but take it in the *spirit in which it is given; you may be sure that I am not going to make any sacrifices, and if your work bears out the promise of your letters and your looks and your samples, I can see where I shall enjoy it as much as you will. Indeed, you may be in for considerable disillusion!*

I do not want you to get your heart or mind set on developing this into a profession by which to support yourself. You will remember what I wrote you early in our correspondence on this point. It is of the utmost importance that you finish your education in all ways as if you couldn't paint a red daub. I fancy you can't help painting birds, and I don't want you to try. But I want you to be an amateur for a long long time, for maybe you would, and maybe you would not, be able to support yourself, even with all the help I could give you both as to developing and then placing your work. I know pretty thoroughly the demands —and also the struggles of young ambition to keep fed. I know you will not misinterpret this advice; if you should, your father will know it is sound.

Yours faithfully

Louis Agassiz Fuertes

Send on whatever you have. Your Cape May [Warbler] record is so far as I know unique. Send it to The Auk, % Witmer Stone, Phila. Academy of Science, Logan Square, Phil., and to Bird Lore.

This, the seventh letter from Fuertes, dealt not with drawings sent him for criticism but rather with the very real possibility that I might go to Ithaca, New York to receive instruction from him personally during the coming summer.

I must have been fairly explicit in asking whether lessons could be arranged for, and the response had been: "So don't embarrass either of us by overvaluing" my instruction, "but take it in the spirit in which it is given . . . ." Never in my life had I come face to face with this sort of generosity outside my own family circle. It moved me deeply. What could I say? What words would make clear how I felt?

The forthright statements that Fuertes made on the finishing of my education were timely. I had been doing well enough in college, though certain required courses had seemed a waste of time. "The real point in studying Algebra," my peers had said, "is that it obliges you to improve your mind." And what I had of mind would go wandering off into the lovely woods, asking itself such questions as, "And when you do find yourself greatly improved, *then* what do you do?"

Fuertes was justifiably concerned over the possibility that I might not be able to support myself and family drawing birds. He wanted me to realize, and fully, how small my income might be at first. He wanted me to know that if I intended to become a professional painter of birds, I would have to be a good one.

Then, in a postscript, he mentioned the Cape May

Warbler that I had drawn—not the drawing this time, but the bird itself. The capture of a Cape May Warbler so far north of where it should have been in December had so interested him that he urged me to prepare a note on it for publication in *The Auk*, the prestigious journal of the American Ornithologists' Union. I wrote the note, a very short one, sent it to Witmer Stone, and actually saw it in print before the year's end (Sutton, 1916, *Auk*, 33: 203). This was, in a very real sense of the words, my first contribution to Ornithology. I was still pretty young, but I was no longer the "young observer" that *Bird-Lore* had called me.

I did not, however, become an associate member of the American Ornithologists' Union until 1919.

In mid-May my mother decided that she should write Fuertes, thanking him for his interest and generosity and (I suppose) expressing her hope that I would behave myself, not be too much in the way, etc., while receiving instruction from him. Mother did not show me her letter, but I know it was a good one, for she expressed herself well. The envelope containing the following reply was postmarked May 24. The letter itself was not dated.

LOUIS AGASSIZ FUERTES
CORNELL HEIGHTS, ITHACA, N.Y.

[May 24, 1916]

My dear Mrs. Sutton—

I cannot tell you how pleased I am to have your fine letter. I only fear two things; one that you entirely overestimate what I have told George I would help him with, and the other that what I can do for him may come a long way short of paying for the sacrifice of money and time that he is making in order to be with me.

As for the first, it is no trouble at all to show him short cuts to the results he has in mind; I shall be painting every day anyway, and I can show him in ten minutes what I couldn't tell him in a week—and he can practice by himself. I am, truly, only too glad to pass on a little of the great and unselfish kindness that fell to me when I was an eager youngster, barking my own shins on my limitations! You must not think that I can make a painter out of George in two or three weeks, anyway: I can clear out some difficulties of technique and perhaps enlarge his vision a little. But the real growth and development will come only as he himself grows and broadens into the field. I myself have had an unusual opportunity to study, as I have been afield in new regions, & have thus become familiar at first hand with most of the birds of this continent north of the equator, and it is primarily due to that, I think, that I have a perhaps unusually broad view of what birds are like. This George would have to work into as he can.

But I think he has the real longing, and it is for this reason that I risk his possible disappointment, as against his possible gratification. I have gone all through the same discouraging inadequacy to get the charm of birds on paper—and suppose I shall die without getting it. But the lure is, fortunately, as persistent as the failure, and success is relative, anyway.

Now I fear I shall have to dampen the party a little. But only a little. We shall not be going down to Sheldrake until the latter part of June, or perhaps July first, on account of school & college, both of which complicate our family plans. Our house here is entirely full of Fuerteses, so we couldn't very well take George until probably about July 1. This situation may alter, in which case I will let you know. And the second untoward & unforeseen thing is that a rather pesky cough, that my little Mary—7½ has been having, has apparently turned into whooping cough—which, if George hasn't had, he shouldn't expose himself to. If he has, it is of course no longer a consideration. The Doctor just decided yesterday that the cough was Wh.-C., and I think he is right beyond doubt.

I went down to Sheldrake on Sunday and have learned that the lowest price by the week at the hotel is $12.00. There is a very nice house, nearer even than the hotel, where George could get a better room, & I fancy for less than the hotel, but they couldn't board him too, and the hotel might ask enough just for board *to make it of no advantage. I will try to find out what we can do in this way.*

When he comes, have him buy his ticket to Sheldrake and I will meet him when he arrives. Sheldrake is a R. R. station on the Lehigh Valley, but the sta. is 3½ miles from the point, where we are, & it is generally easy to pick a ride down with an auto-owning friend. Sheldrake is halfway between Geneva and Ithaca. Coming from Buffalo, the Black Diamond Express will take him to Geneva, & a local train the remaining 20 miles, if he gets to Buffalo early. B. D. leaves 8 A. M.

Sometimes we have a steamer service on the lake, which is much the nicest way to get down.

I want to tell you that I have enjoyed George's letters and his enthusiasm from the very first and I know he is the kind of a boy we shall all like, from Mary up. If the sacrifice is not too great, I am sure the vacation & change will be good for him, irrespective of what advance he can make, and we are all anxious to meet him. We too love music and have all we can, tho' none of us performs but my mother (77), who stills plays gloriously.

Faithfully yours,

Louis A. Fuertes

LOUIS AGASSIZ FUERTES
CORNELL HEIGHTS, ITHACA, N.Y.

May 15, 1916

My dear George—

I have delayed even longer than usual this time in answering your letter. This time I have a better excuse, because I have been much of the time away from home, and exceedingly busy when I have been here, trying to catch up.

I am writing now to say that my summer plans are about crystallized, and we shall be at Sheldrake, 20 miles down Cayuga Lake from here, and if your plans work out so as to make it possible for you to come this way, I shall be very glad indeed to see you and help you all I can.

The cottage we have is little, and we fill it up ourselves. So I can't offer you an invitation to stay <u>*with*</u> *us. But there is a very nice little country hotel, where we always have our meals, and it is only two hundred yards or so from our cottage. Its rates are very reasonable, and you could comfortably settle yourself there, with a fairly good light, and pass the time pleasantly and perhaps profitably, for as long a time as you like it, or can arrange to. I can't tell you the exact rates, but probably $10 a week wouldn't be far from what it would cost. I shall be painting, myself, most of the summer, probably, and will be glad to help you, or have you see how I go about it—the* *tools I use, etc. etc. any time from early July on, if you find it advisable to come on. I should of course not want you to go to any unnecessary expense: I know all about such things, as my own youth was exactly (and strangely) similar to yours: my own father was a college professor, so I know about the financial limitations. Your work here would cost you just your fare back & forth from Bethany to Sheldrake and the hotel charge as long as you stay, and as far as I can see that is all, unless you want to buy some materials to work with here in Ithaca.*

If the idea appeals to you as practicable, just let me know, and I will get such information, as to hotel rate, etc. as you wish, on receiving your reply. As far as my own part of it is concerned, do not fear that I shall greatly obligate you! I am working anyway, and I have no patent on anything, so you are welcome to learn all you can by watching me work, and I should greatly enjoy knowing you "in the flesh," as naturalists call it. I shall promise beforehand not to put myself out greatly for you.

It is a quiet little country place on the lake shore, with no fashionable people to dress up for or other intimidating circumstances! I pass most of the summer in a soft shirt and khakis, & my kids live mostly in their bathing suits. So you needn't fuss about clothes or other things. For myself, life seems too short anyway to encumber the hot season with artificialities, and we are all agreed on that score.

Put this thing before your father, unless your plans are all made for other means of utilizing the summer vacation. I shall not try to make it too attractive to you, but will simply say that it is a good time and place to work; you can probably make considerable progress, and that I shall myself greatly enjoy knowing you better. There are many forms of mild summer enjoyment—we picnic a good deal—the swimming in the lake after July 4 is very fine, there is amateur tennis in the region, birds are abundant, and I have a canoe & the fishing is sometimes good—at least for perch! Most of our amusements include the kids: we have 2, and some 15 more are harbored near, tho' it is not a real summer colony.

Very sincerely your friend,

Louis Agassiz Fuertes

The air is full of birdsongs now, and the trees & woods sifted full of warblers, etc. Big time is at its height.

Well before the important letter of May 15 reached me, I had decided to save every penny that came my way for the trip to New York. I had been making precious little as a janitor, and the season for lawn-mowing, garden-tending, and silo-filling had not arrived, so I applied for work on the county road that connected Bethany with the city of Wellsburg, on the

Ohio River several miles away. The pay was good for those times, about $3.50 per eight-hour day if I remember aright. The road-boss seemed glad to give me the job, for he was short of men.

The road was being surfaced with red brick. A sloping stretch not far south of town was receiving attention. Here ditches were being dug, a stone-crusher was at work, and bricks by the thousand were being unloaded, stacked in piles, and laid between narrow strips of cement. My chief job was unloading and stacking bricks, but I dug ditch and fed the stone-crusher as the need arose. It was the toughest work I had ever done and to my surprise I enjoyed it. There was something almost romantic about finding a shady spot for the sack-lunch that I ate with the other workers; about swapping yarns with them; about realizing, as I walked home, that I was grimy with dirt. Fortunately, the Suttons had by this time moved to a two-storied house in which there was a good bathroom.

In the back of my mind all the while was Fuertes, the train-trip that I was about to make, the pictures I would draw, the reputation these eventually would make for me.

The road-boss who made the rounds never told me in so many words that I was in charge of the gang of six or seven men with whom I worked, but what he said and the way the gang acted made perfectly clear that it was I who was being held responsible for the neatness of the brick piles, the depth of the ditches, etc. This realization that I was in charge of something made me feel important. On my way home, evening after evening, I felt that I had earned every cent that I had made that day.

One brawny young black who was in my gang made himself unforgettable by nicknaming me "Chicago Town." Where the name came from goodness knows, but he used it constantly, especially when reminding me, in a cheerful, not patently obsequious voice, that we were working too hard. "Not so fast, Chicago Town," he would say. "Not so awful fast. We gets all the work done all right, all right, but not so fast. We likes it better that-a-way, and you likes it better that-a-way. Ain't that the truth, now, Chicago Town?"

The bricks piled up day after day. So did the money—for the trip to New York and Fuertes.

LOUIS AGASSIZ FUERTES
CORNELL HEIGHTS, ITHACA, N.Y.

May 26, 1916

Dear George—

I waited until the pictures came before replying to your good letter, and I want to tell you right uway that they are far and away the best you have done at all, and they show a kind of improvement quite aside from the things I have tried to help you with, and of a kind that must necessarily come from inside yourself. Your little goldfinch is a real goldfinch—just as your wood pewee is a real wood pewee. They are very much better painted than any you have done before, and you should be very much encouraged. If I can help you by really watching you work, & having you see how I go about it to get what I am after, you ought really to make the summer a good investment, for you have worked long enough & hard enough to have overcome yourself a great many technical puzzles, and I see a confidence in your handling that was far from evident in your earlier work. These are not easy subjects—any of them—and you have done finely with them. I shall not go into detailed criticism now, but will hold them for a first lesson when you come on.

I get most of my supplies right here, and it will be best for you to work for the few weeks intervening with the material you have, and I will look them over, and see what you really need. I myself use very simple tools, and can probably lend you everything until we have a chance to come to town and buy out the stores!

This Sutton drawing of an Eastern Wood Pewee *(Contopus virens)* is the one that Fuertes called "a real Wood Pewee" in one of his letters. The date at the bottom is that on which the specimen was collected at Ypsilanti, Michigan, not that on which the drawing was made.

Wood Pewee, ♀
Sept. 15, 1913, Ypsilanti, Mich.

I can help you on paper—that is one of the great problems, & I think I've got it all solved: a paper that will work equally well with a thin wash, full brush or almost dry, and hold the color after it's once applied and dry. Sheldrake is a way station on the Lehigh Valley, halfway between Geneva and Ithaca. If you get into Buffalo before 8 A.M. there's a fine train (Black Diamond Ex.) that goes to Geneva with only a couple of stops between, & connects well with a local that will bring you to Sheldrake in about 40 minutes. If you miss that there's another that leaves about 11, stopping for milk, etc. all the way from Buf. to Sheldrake. Ithaca is 20 miles further along, so just buy to Sheldrake. I'll give you full details later.

I don't know just when I can get the information we want concerning the other schemes—rooms outside the hotel, etc. You will really need a room with a good light, to get the value of your stay, but we'll get the best we can and make your money go as far as it possibly can. It's a very simple place, and I am sure the people can be interested in our favor. It's a very "birdy" place, & occasionally something really interesting turns up.

Yours sincerely

Louis A. Fuertes

If you like kids you'll have all you can do to dodge them: I think there are about 20 on the point, from 3 to 16, mostly cousins, who swarm all over anybody that gives them a chance! But they're nice kids, & we have a lot of fun, for all ages merge in the summer-time!

If I had been having doubts about my progress as a bird painter they were dispelled by this, the ninth letter that I received from Fuertes. Here he mentioned an American Goldfinch drawing that I cannot remember and an Eastern Wood Pewee drawing that for some reason I kept—and still have. The pewee was done not from a specimen I had just shot, but from a prepared skin that I had found in a drawer in the Biology Laboratory in Oglebay Hall, a specimen that Professor Milton Hover had brought with him when he had moved to Bethany from Ypsilanti, Michigan. When I look at this drawing today it seems woefully stiff and amateurish, but it must have been better in certain ways than earlier sketches I had sent for criticism, for Fuertes went so far as to call it a "*real* wood pewee," by which he was trying to say, I believe, that the bird was perched in a characteristic attitude and that it had a three-dimensional quality that he had been wanting my birds to have.

So Fuertes was actually expecting me to be with him during the coming summer! "These are not easy subjects . . . and you have done *finely* with them," the letter read. "I shall not go into detailed criticism now, but will hold them for a first lesson when you come on."

A "first lesson" with the great, the one and only Louis Agassiz Fuertes: how could any such thing possibly come to pass?

LOUIS AGASSIZ FUERTES
CORNELL HEIGHTS, ITHACA, N.Y.

June 12, 1916

My dear George—

Your letter has just come. I haven't been able, since I wrote you last, to get down to Sheldrake. Indeed, I've been away "on the road," for a good part of the time, lecturing around about bird songs!

I have myself thought of a tent as the solution of your problem, and as far as sleeping quarters go it would serve perfectly. There is ample room for a whole circus on the point, and we could fix you up very snugly, and probably find a tent for you. But your principal need while here will be a light, quiet, comfortable place to work, which a tent would hardly furnish.

But I am quite sure that we could fix a place for you in one of the first floor rooms we only use occasionally, where perhaps we could both paint mornings. I shall probably have quite a bit of work early in the summer, and as good a way as any for you to learn to overcome such technical difficulties as you mention is to watch somebody else, & then go at it yourself. It's like making a birdskin: speed insures good results, once your nervousness is conquered.

You just plan to come on, any time after June 28 or July 1 (by which time we will be about established) and we will make every effort to reduce expenses and improve facilities, so that you can stay as long as possible. Perhaps you had best knock off painting now for a time, & rest your productive cells: it is hard coming fresh again after a real stale streak, which I suspect you are now fighting. If you come to me fresh, it will be easy to shoot you right ahead. Otherwise you may easily have the benefit of what I can show you. Bring along all your "tools" and whatever you happen to have done, and let me know on what train you leave Buffalo. I'll meet you at Sheldrake. Remember to change at Geneva if you're on a fast train, or you'll land in Ithaca, 20 miles too far.

I have enquired of a neighbor concerning a tent, and find that I can easily get for you an excellent tent, with fly—8 x 10. This is big enough to make you very comfortable, and might work out even as a studio. So I think you can confidently count on making your money go only for board, which is both good and reasonable where we all eat. I believe that $7.00 a week will cover your eating—it is barely possible that when the time comes, and you can tell about how long you can stay that Mr. Reese (the propri.) will make a special rate for you.

If you have a tent of your own you are used to,

or prefer, bring it along: the one I have in mind here belongs to a friend, who may possibly want to use it, tho' I think not. I have a small 7 x 7 wall tent, but it is hardly tight, being quite old, and has no fly.

We can surely fix it so that you won't have to rent a room, & can have a good place in which to work.

As to finishing an unsatisfactory start—I know of no greater waste of energy and patience, & no surer way to go stale. The greatest lesson I ever learned was the cheerful destruction of a false start. Don't be afraid of your rubber: use it all the time & keep it always handy, & don't ever go ahead on a picture till the <u>drawing</u> satisfies you fully. If necessary, carry it farther than you need for an outline merely,—suggest shading enough to show if it is going to "solidify" right, & then erase the shading before painting. But a thing falsely drawn will never prove up, & you'll waste a heap of time, and be going backward all the time, trying to convince yourself you're righting a wrong. Better far to build well, as you go along, and remove every fault as it comes. <u>That</u> is all progress, and the results will show immediately.

I have meant to write and thank your mother for her very nice letter to me: I've had such a multiplicity of things to do that I am not now sure whether I acknowledged it or not. Will you tell her that I was made very happy to get it, and that she need have no fear of your being an inconvenience or trouble to us in the least? We live extremely simply and quietly, with no "life" anywhere around, and will be delighted to have some youth and enthusiasm around.

Yours as ever,

Louis A. Fuertes

By far the most important part of this, the tenth letter to me from Fuertes, was the next-to-last paragraph. But the rest of it had me so excited I could hardly sleep at night. That penultimate paragraph about use of the eraser should be memorized by every person who hopes to draw birds well. The words "A thing falsely drawn will never prove up" might well be a motto for Everyman's wall, for philosophically the statement applies to so much more than the mere delineation of a bird.

Throughout virtually the whole of my career, I have used a well-sharpened hard pencil in drawing the outline of my bird before putting on any paint. I have learned that all too frequently there is such a thing as thinking that the outline is right, then charging ahead with adding details, only to discover that the whole bird is over-size. Even today, I tend to make a bird too big if I intend it to be life-size. A bird too big in outline may be ever so attractive when colored-in, but it will still be too big.

In southwestern Mexico I recently had an opportunity to draw a Cinnamon Hummingbird from a specimen found freshly dead under a window it had struck. I drew the little thing in detail, found that I had drawn it too big, erased it and drew again, only to find that it was still too big—half an inch too long, in fact, which is a lot when the whole bird's length is only four inches! Finally I settled for laying the tiny corpse on the paper, putting a pencil mark where its bill started and another where its tail ended, and forced myself to stay within those bounds. Meanwhile I had been obliged to use more eraser than pencil!

A sweet little girl named Miki Lara, who sometimes watched me at work, was so impressed with the time I spent erasing that she came to consider the eraser every bit as important as the pencil. She learned the word *eraser*, decided to make some drawings of her own, and called the attention of passersby to the fact that she was making a drawing *with her eraser*. How Fuertes would have enjoyed the innocence of that remark!

The shape of a bird's bill is sometimes hard to get just right in a drawing—the curves of its contours, the way the upper mandible covers the lower at the tip in some species, the "tooth" on the upper mandible in shrikes, the angle of the gonys (lower edge) in gulls and terns, etc. In drawing a Long-tailed Hermit Hummingbird, again in southwestern Mexico, I found that the slender bill's curvature was extremely difficult to draw. The bill did not curve from base to tip; it started out straight, stayed straight for about a quarter of its length, then curved. I must have drawn the bill a dozen times before I felt that I had caught its curvature correctly. The final draft had been accomplished chiefly through wise use of the eraser!

LOUIS AGASSIZ FUERTES
CORNELL HEIGHTS, ITHACA, N.Y.

June 25, '16

Dear George—

Don't worry about anything. I've got a good tent, and a good air mattress. There's plenty of lumber down there, and I think we can floor your whole tent for you. All you need bring is your clothes and your painting things and blankets & sheets. It's all fixed so that all your money can go for board, which we will get down as low as possible on a weekly or monthly basis, to keep you as long as possible.

If you get to Pittsburgh so as to get a night train to Buffalo, you can leave B. at 9:30 A.M. of the 29th, on the Black Diamond Express over the Lehigh Valley. Get off at Geneva at 11:42, and take a local train from the same station at 12:20, arriving at Sheldrake at 12:43. There will be a rig there that will bring you and your stuff down to the point. I shall expect you on that train unless I hear from you to the contrary.

You can't get out of Buffalo again until 6:45 P.M., on a local all the way, which will let you off at Sheldrake at about 10 P.M.; this would bring you to the point not earlier than 11, & it would be much better if you can get the Diamond *at 9:30, which is easily done on the night train from Pittsburgh, if you come that way.*

We leave here on the morning of the 29th, getting to the point about 10 or 11—a little before you do. We'll wait lunch, if necessary, for you but probably I won't be able to meet you at the station as I shall be much needed in settling the cottages, as I'm the only husky man in the outfit. Don't worry about being a trouble—I never take trouble with anything. So if you can make the connection on the 29th, I shall simply expect you on the 12:43 Local, & you needn't bother to let me know. You'll have to spend a night somewhere on the journey anyway, and the sleeper to Pittsburgh is about as cheap as any hotel you'd be safe in, & you gain a whole state's width in distance. If you can't come till next day you might telegraph me at Sheldrake, & the message will be phoned to me.

I want if possible to get this into the Sunday P.M. Collection, so you'll lose no time in its delivery.

Yours in anticipation

Louis A. Fuertes

So I was not to worry about anything! I'd be living in a tent at Sheldrake Point and that would save me so much money that I'd be able to buy some of those expensive brushes that Fuertes had mentioned, and some paper of exactly the right sort, and perhaps some tubes or pans of paint, including one of lamp black and one of Chinese white.

This time I talked plans over very thoroughly with my parents. My one "good suit" I would wear on the train. The one suitcase available that was large enough for blanket, sheets, and clothing was old, indeed it was not very presentable, but it would have to do. My parents were not to help further with finances. My earning and saving of money had been almost ostentatious for weeks. The whole village knew about it. I could almost hear people saying, "That Sutton boy who's so much interested in birds is going *to New York* for the rest of the summer."

The trip from West Virginia to upstate New York was the longest by far that I'd ever taken by myself. Dressed in that one "good suit," whose greenness matched pretty well that of its wearer, I economized by eating home-made sandwiches for supper on the train, sleeping in coach rather than Pullman on the night ride from Pittsburgh to Buffalo, and breakfasting on salted peanuts. I simply gloried in refusing to spend money. The refusal was almost religious.

When I arrived at Sheldrake Springs, Fuertes was not at the station to meet me. For a long moment I felt lost; but a man who knew where the Fuertes house was told me which road to take for "The Point," so off I went on foot. The day was pleasant, so I did not mind the long walk. When I had reached the cluster of buildings at "The Point" a truck drove past, slowed almost to a stop, and dropped off a man who waved thanks to the driver as he called out, "George, I'm glad you're here. The place is yours. Just do anything you darn please, now, while you're here!" It was Fuertes.

Only a few yards from his house, which was close to the lake's edge, was the tent that Fuertes himself had put up for me. On a folding cot was an air mattress, the first I had ever seen. There was a camp stool across from the cot, a small table, a wash-basin and soap dish—all the comforts of home. During that first night I was wakened by the sound of something being dragged across the wooden floor. A rodent of some sort had found the soap under the cot and was making off with it. I rescued it just in time.

I ate with the Fuertes family at a hostelry in the village. Meals were festive. Fuertes was at work on two sets of watercolor illustrations—a big series of mammals for the *National Geographic* magazine and nine drawings of common birds for a book titled *The Way to Study Birds,* by John Dryden Kuser. When Fuertes was at work he slaved. Nothing diverted him then. But when mealtime came, he sought release from all that had driven him. Release took various forms—funny stories, many of which included quotations in dialect; frank comments about people in the public eye; pranks meriting some such phrase as hi-jinktumgee. Alphabet soup, welcomed with a mild uproar, gave us all a chance to compose cablegrams spelled out on lettuce leaves.

There was no telling, simply no telling, how that irrepressible fun-streak in Fuertes would pop out. After our first four or five meals together, I was prepared to believe the story that I'd heard about him and the oh-so-proper waiter who needed to be reminded that very nice people can be highly informal on occasion. According to the story, the restaurant was fancy. Conversation at the table had ebbed. The disagreeably proper waiter was close by, where he could not fail to hear every word. Said Fuertes, "I'll bet I'm the only person in this whole place who has a dead mole in his pocket!" Everyone at the table knew full well that their beloved Louis was capable of having precisely what he'd said he had precisely where he'd said he had it; but the waiter, clinging desperately to belief that such a statement could be nothing but bluff, stood by as if transfixed. So out the lovely little specimen came, its silken fur gleaming in the candle light, to be placed on the table where all could see it. Everyone laughed merrily, even a bit boisterously. Everyone but the waiter.

The Fuertes family awed me, not only the famous father, but the other three as well. Mrs. Fuertes, whose first name was Margaret (shortened to Madge by husband and close friends), told stories extremely well, and her opinions on any subject were worth listening to, but her special gift was that of leading conversation toward a point at which a yarn from her husband would be appropriate. Mary, the daughter, a little girl of great charm, reminded me of flaxen-haired princesses I'd read about in fairy tales. Sumner, a bit younger than Mary, and nicknamed Shub, was elegant even as a youngster, and handsome.

The small, well-lighted room in which Fuertes worked was at one corner of the house. Here he had a table, drawing-boards of various sizes, a vase full of watercolor brushes, and tubes of paint, little porcelain "pans" of paint, and paint boxes galore. Lying on the table was the beautifully prepared skin of an Eastern Phoebe—the bird he was about to draw. When I looked closely at the specimen I realized that it was no average "scientific skin." Its back was not flat, but rounded; its feet stuck up; its feathers had a fluffy, recently preened appearance. All this was no accident, for Fuertes had prepared the specimen himself, to serve as a model.

Fuertes let me watch him at work. At the time this did not strike me as remarkable. I was eager to learn. I had traveled a long way. What I didn't realize was that Fuertes might have been wretchedly self-conscious, even tense, as I stood there watching. The task at hand for him was finishing an acceptable drawing, one to be reproduced in color. If the original, when completed, proved to be "off" in any important way, it would have to be done over. Having a worshipful youngster at his elbow might have been ever so flattering to his ego, but it also divided his attention. Flowing-on a watercolor wash is extremely tricky; this I know now from years of experience. I had no idea of it then. When Fuertes put the background wash on that phoebe drawing he was working toward two ends, serving two masters. He was creating part of what was presently to be a bird portrait of immense charm; he was also showing a pupil just what to do and, even more important perhaps, what not to do.

The lightly penciled-in drawing was firmly attached to the drawing-board with thumbtacks. The bill of the bird in the drawing was slightly open. I thought at first that this was to indicate that the bird was singing. Not so. Fuertes picked his model up, looked carefully at its bill, examined his drawing, and stood up, saying, "We've got to catch a moth. Come on, help me find one!" So out we went and around we pranced until a moth of proper size and beauty had been flushed from the grass and captured. I watched Fuertes pencil-in that lovely lepidopteran held so tightly in the phoebe's bill. Who would ever know that I, not the phoebe, had caught it?

The background wash, pale grayish blue in the upper third of the picture, light grayish green in the lower two-thirds, was to go on before any part of the bird or its perch received color. Fuertes took from the vase a big brush, mixed some blue in a saucer, added water, strengthened the tone by adding more paint, tested the wash on a piece of white scratch-paper, poured in more water, then rinsed the brush thoroughly before mixing the green wash. The two washes were in separate saucers. "Be sure to have plenty," advised Fuertes. "And do all the mixing and all the testing before you start."

Now, tilting the drawing-board so that the wash would flow downward, he moved the paint-laden brush rapidly from left to right across the top of the paper. "Never let the lower edge get dry. You have to work fast!" I could see the color from the first stroke flowing down into the upper part of the second stroke. No edge formed. Presently the whole top third of the paper was pale grayish blue—all but the phoebe's head, up to the very edges of which the brush had moved swiftly, unerringly. Fuertes flicked the brush to rid it of blue, rinsed it, flicked it again, and started with the green, all this before the blue had had a chance to dry. This time the stroke was not strictly horizontal, for the uppermost part of it was to represent distant treetops. Nowhere did a hard edge form. The two tones flowed together smoothly, harmoniously, at treetop level. "Before you make another move," cautioned Fuertes, "that background wash has to dry. If you touch it while it's damp, its smoothness will be ruined. It's best to forget the whole thing for a while." Whereupon he led me outdoors again, this time heading toward "The Point," which was only a short way from the house. A dark-backed, short-tailed swallow flew past us at eye-level. "Know what that bird is?" queried Fuertes. "A Rough-wing," I replied. "Good boy!" said Fuertes. And my day was made.

When we returned to the house, the background wash had dried. Now was Fuertes to bewilder me by putting a lot of seemingly unrelated dark lines and spots here and there within the area that was to become the phoebe. These were what he called "the deepest darks," the very phrase he had used in his letters. The brush he used was not small, but it had a fine tip. After he had put in a few very dark marks, he added less dark lines and patches, all neutral in tone. The phoebe by this time had a decidedly leprous appearance, but I could see that the wing was beginning to take shape.

Matching colors and ascertaining that these would, when completely dry, represent the bird's plumage accurately, took time. I watched the test-dabs of paint dry on the scratch-paper. Presently the brush settled down to work. The phoebe's whole head, pulled together by deft strokes of dark paint on its upper half, became feathery. To accent a highlight, Fuertes "pointed" the brush with his lips, touched its tip to part of the fluffy crest, and lo, there came into being, all at one instant, a curved surface, a softness, an effect of light and shadow, a three-dimensional quality. I had seen it all happen—right there within touching distance.

So engrossed was the master by this time that he said little. He did not need to say anything. As the brush, now carrying an olive-gray tone, moved over part of the wing-area, the middle and greater coverts overlapped convincingly. An important fact became apparent: not every blessed feather needed to be painted separately; to delineate each separately in great detail would be a mistake, for such a handling might result in scaliness rather than featheriness.

A big blue fly became a nuisance. Not content with

drinking at a saucer, it blundered about, at times between the artist's face and his picture. "Gosh," said Fuertes, "I wish somebody'd kill that brute!"

The fly buzzed drunkenly about, alighting on the table, ceiling, and floor, at length on Fuertes's head. In a vain attempt to snatch the unwelcome insect, Fuertes swung with his right hand, sending his brush to a corner of the room. I retrieved the brush. The fly continued its moaning. As it moved past me I watched closely, took aim, and, forgetful of everything but the need for eliminating "that brute," whacked my hands together. The noise was frightful. Fuertes jumped. Off went the brush again as the drawing-board clattered to the floor. I was speechless with dismay. Fortunately the drawing was not damaged. As Fuertes resumed work he shouted, this time almost fiercely, "Good work, boy! You winged him that time, didn't you!" I'm not sure that I even grunted a reply. The fly had been more than winged; it was entirely dead.

Fuertes painted the moth with great care. I marveled that the brush could capture the intricacy of the wings' markings. After the moth came the perch, then a bit of wild morning-glory vine, finally the phoebe's eye. Suddenly the picture came to life. The eye twinkled. The soft plumage rose and sank. I had to look twice to be sure the moth was not disappearing down the phoebe's throat.

What I have just said sounds exaggerative, I know. Remember, please, that I am reporting on how I myself felt when I beheld that finished portrait. Not on how someone else might have felt or should have felt; but on how *I* felt. That drawing, the drawing itself, had life: there is no more honest way to describe the effect that that picture had on me.

The brushes I had brought with me were all small; some were tiny. They would not do. To Ithaca we went, where I bought some fine new ones. Not many, for they were expensive. We visited Fuertes's real studio that day, too, a separate building back of the Fuertes residence at the corner of Thurston and Wyckoff avenues. A double window on the north side of the big room was large; along the opposite wall were cases holding a wonderful collection of birdskins.

Most of the specimens that I looked at that day, were of species I had never seen before. Especially lovely was a Swallow-tailed Kite wrapped in white tissue paper. Fuertes handled this specimen almost reverently, for he was determined not to rub the "bloom" from the dark plumage of its upperparts. The Whip-poor-wills, Chuck-will's-widows, and other nightjars fascinated me. One of these I examined too roughly. "Careful, boy!" Fuertes warned, "those birds are delicate!" As I returned the specimen to its place in the drawer I did so thoughtfully. I had learned an important lesson.

Fuertes showed me books with colorplates by Archibald Thorburn, a British artist I'd never heard of. "Some of those Europeans have done fine work. The feathers of Thorburn's birds are really soft. I like his stuff!" In the 1913 *Annual of Swedish Art*, a handsome publication, were colored reproductions of oils by Bruno Liljefors, again a name new to me. "Every Liljefors painting deserves close study," said Fuertes. "His birds and mammals are part of a habitat. He is just as deeply interested in their habitat as he is in them."

It is difficult to account for the fact that I did not draw a single bird while I was at Sheldrake. Parts of birds, yes, but never a whole bird. I believe I was inhibited, afraid to face the crudeness of my efforts. I would watch Fuertes every day, often for hours at a stretch. Then I would tackle something on my own. My sketch of a kingfisher's head with bill pointed straight at me was a total flop. Drawings of weeds were better, though one that I still have is woefully "tight."

I enjoyed walking along the lake shore. The difference between a dry pebble and that same pebble as seen under water fascinated me. I loved the way reflected sky robbed incoming wavelets of their transparency, sometimes flashing brightly enough to make me blink. Among the pebbles at the water's edge I occasionally found fossils. Some of the prettiest of these I saved.

A Spotted Sandpiper that I found dead I decided to mount for my collection. This work took the greater part of a day and Fuertes may have felt that I was misusing my time, though he voiced no objection. "When you get home," said he, "just put that bird on the ground near the creek and study it carefully. Remember what I've said about light, shadow, and reflected color. Black feathers struck by sunlight aren't

Sutton drawing of a weed made at Sheldrake Point on the west shore of Lake Cayuga in the summer of 1916 while Sutton was with Fuertes.

really black. You know they're black, of course, but blackness is not what you see. The only really black spots on your sandpiper's underparts are those that are in deep shadow. Black spots struck by sunlight are gray, not black, and the gray may have a blue tone if the sky is blue or a green tone if there's green vegetation close by." Never have I made a bird drawing since that day without bearing in mind those important statements.

The greatness of Fuertes came out in unexpected ways. On an automobile trip to the Montezuma Marshes at the north end of Lake Cayuga, we stopped at the home of a man named Foster Parker, who was to accompany us. Mr. Parker was rearing a penful of Wood Ducks. Big tubs in the yard were alive with tiny frogs, food for the growing ducklings. As we drove along the edge of the wetlands, many birds flew up from the cattails, among them an occasional Marsh Hawk. "Look, there's another Marsh Hawk, a brown one this time," said Fuertes to his daughter, Mary. "That's a bittern," corrected Mr. Parker. "Right," responded Fuertes cordially, "I should have known better." Turning to me he said, *sotto voce*, "Foster Parker knows these marshes and their birds better than anyone! I'd give a lot to know them half as well as he does." I have never forgotten the sincere modesty of that statement.

Mary and her father were very close. I recall the look on the little girl's face when Fuertes called her attention to the singing of a Yellow-throated Vireo. "Just listen to him," said Fuertes. "What he's singing is 'Mary, Mary, come 'ere'!" Mary would listen, and smile, and the little bird, high in its tree, would wait a while before singing again its low-voiced, ever so slightly raspy, six-syllabled song.

Mary Fuertes was gentle, friendly, and, even as a little girl, scholarly. She appealed to me deeply. It was she who told me of the Athrabrians, a secret society to which she and her brother belonged. Where they had found this impressive name I never learned. I did not expect to be invited to a meeting, for I knew that Athrabrian activities were shrouded in mystery. I was, however, to see one of their most-prized possessions, a wave-washed letter found along the shore, a tragic missive, the only legible part of which read: "Dear Sadie, I told you I was happy. I am not. My God, to think of him behind them iron bars."

Day after day I watched and learned from Fuertes. I was not with him all the time, but usually I was within calling distance. His drawing of a beaver—for the *National Geographic*—I saw him change drastically when it was almost finished. The part that Fuertes didn't like, all of the animal except its head, he scrubbed out fiercely with a sponge. I was aghast when I beheld that wreck of a drawing. But Fuertes, powered by conviction, did what he felt he had to do, and he won. His "new" beaver was among the most effective drawings of the series.

Before I left Sheldrake Springs, Sumner Fuertes gave me a Lake Cayuga "lucky stone," a little ring of smooth gray stone about three quarters of an inch in diameter. This I strung on my watch chain and wore for years. Fuertes himself gave me one of his old paint boxes—a treasure I was to take with me four years later down the whole of the Labrador coast; seven years later down the Abitibi River to James Bay; ten years later down the Missinaibi River and the east coast of Hudson Bay; eventually time after time to the New World arctic, repeatedly to Mexico, once to Iceland. Never since the summer of 1916 have I painted a bird without having that battered old paint box somewhere close by.

When I returned to West Virginia from "The Point" in the late summer of 1916, the family was living not in Bethany but in Pughtown, a village about the size of Bethany and, like Bethany, back several miles from the Ohio River in the state's Northern Panhandle. The closest sizeable town was New Cumberland. Father was a "supply preacher" at the Congregational Church in Pughtown.

I had with me in Pughtown some good watercolor paper, three fine new brushes, and, most important of all, the old paint box that Fuertes had given me. There was something mystical about that paint box. Having it near me warmed my heart and gave me confidence. Finding a table and a window where the light was good, I set to work drawing—first a Chimney Swift that someone had found dead, then a Yellow-throated Vireo, a Least Flycatcher, and a Black-billed Cuckoo that I had shot near town. I was not especially pleased with the drawings. Indeed, they seemed poor. They seemed to furnish proof that I was not properly analyzing shadow color. I resolved to do the best I could, to keep trying, but to send nothing more to Fuertes for criticism unless it showed real improvement.

The sketches I had made at "The Point" were of special value to me now. I studied them often, trying to decide just how they were better than earlier ones. Drawings of weeds that I had made seemed stiff and fussed-over. They looked almost as if they'd been done with artificial plants as models. But they showed that I'd been paying close attention to blueness of sky reflected from glossy surfaces.

My mother, glad to have me safely home, wrote to Fuertes from Pughtown without my knowing that she was doing so. Here is what Fuertes said in his reply:

LOUIS AGASSIZ FUERTES
CORNELL HEIGHTS, ITHACA, N.Y.

Aug 13, 1916

Dear Mrs. Sutton—

I had been meaning to write you, or George, but since he left I have been letting things go rather easily, as I have had a long hard year that just closed when I finished the work on hand while he was here.

I think George has a very sincere interest, and a great deal of promise. Whether or not he has genius or talent in any high degree it is yet too early in his life to determine. I talked very plainly to him of his work, his prospects and his policy for the next few years, and I think he knows what I think, if it is of any value to him, in that direction. He is in many ways still a boy, and it is very wise, I think, to let him catch up with himself, meanwhile studying both his painting, his birds, and his college work. By all means finish that, with full concentration.

Personally, I can't define to myself genius, as applied to science and art in combination: I think everlasting attention and practise and study is the way to it. I saw marked improvement in George's work while he was here, and tried as well as I could to show him ways over difficulties, & technical resources. His observation he must get for himself, and his correct analytical vision he must perfect also, & it will take time. I was not disappointed with him in any way; I was so tired myself that I was all the time aware of my dullness and inability to express what I tried to, and learned at least that my forte does not lie in teaching.

But if George can remember the things I tried to show him, and practice & perfect his abstract painting ability, he will at least have profited by his long and expensive journey.

I cannot tell you how much I appreciate your

letter, although you have entirely magnified what I did. George was no trouble at all, and I was very glad indeed to have him. He fitted in perfectly with the life here, and left many friends behind him. He must continue to send me his studies, and I will be glad to give him criticisms. His improvement should begin now, rather than while he was here, as I had to crowd the ideas in pretty fast, and they should take root & bear fruit with his subsequent work.

Please thank Mr. Sutton, also, for his letter when George came on. I meant to reply (though he told me not to). You all think much too generously of my help, which was very easy, as far as it went. My best, too, to George.

Faithfully yours,

Louis A. Fuertes

During the fall of 1916 I was on the run. I was now a sophomore in college. On five days of each week I mopped the floors and tidied the furniture of Oglebay Hall, but I was afield as much as possible. The woods were bright at this time of year, especially the beech trees with their yellow leaves. A few of the birds that I collected I mounted, but most of them I preserved as scientific skins. My interest in Ornithology was known throughout the countryside, so birds were brought to me from time to time, among them a lovely adult drake Hooded Merganser that had been shot along Buffalo Creek. A crippled immature Red-tailed Hawk given to me by a farmer glared at the world with wide-open mouth and stuck-out tongue as I wrapped it in an old towel. Thus immobilized, it served as a model for a direct-from-life drawing of its head. This portrait I considered the best I had ever made, so I wanted Fuertes surely to see it.

Toward the end of November a neighbor named McKinley Tener shot a hen Wild Turkey in the West Virginia mountains near the Cheat River. Remembering my bird collection, he wondered if I would like to keep the skin of this trophy with the understanding that the bird itself be returned to the Teners for dinner on Thanksgiving Day. Needless to say, I wanted the skin.

I had had no experience with Wild Turkeys, so was more than glad for the opportunity to handle this one. The big bird's crop was packed with chestnuts, every one of them whole and perfectly clean. As I removed the skin it occurred to me that Fuertes might want it for his collection. Here was a chance for me to show that I appreciated what this kind man had done for me. In answer to my query about the matter came this reply:

LOUIS AGASSIZ FUERTES
CORNELL HEIGHTS, ITHACA, N.Y.

Dec 2, '16

Dear George—

I was delighted to get your good letter, and am more delighted, if possible, to learn that you have a turkey for me. I think you better make it a skin, with the neck turned to the side, to rest on its belly. I'd love it mounted, but my room is so small, and mounted things collect so much dust. I have no wild turkey and have long been trying to get one.

I got a fine ad ♀ goshawk the other day—weighed 3 lbs less ½ oz., & had been feeding on a barred rock hen!

I have to go clear to St. Louis on Monday for 1 lecture. I'd hoped & planned for another in Chicago, & maybe Cleveland on the way home, but they all fell through. It's a very expensive loss of time, & now I'd give a good deal to get out of it.

I'd be mighty glad to see your hawk pictures. I've had a couple of fine birds, but through one thing or another was prevented from painting them.

My animals appear in the next Geographic, and I've spent most of the fall on the "layouts" and plans for the second series—small mammals— & have a good start on them in their finished form. My studio is full of rats, mice, squirrels, rabbits, skunks, weasels, etc. ad lib.

I'm glad you're busy, and think you are entirely

right to get your liberal education first as a background. The painting will not, I think, suffer greatly, and in the meantime you will be getting an indispensible asset and developing yourself to a point where difficulties are easier to solve and less discouraging.

I shall be glad, always, as you know, to see your study, and will always criticize it for you honestly and carefully. The photos of the hawk pictures will help, & I'm eager to see them tho' of course the color I can't judge so well. I nearly commandeered a fine ad. redtail; only my departure for the A.O.U. in Philadelphia kept me from doing so. It was brought in by a farmer to have mounted down-town.

I haven't had my gun out of its rack all fall, & don't believe now that I will.

Thank you again, and very heartily, for the turkey, which is a valuable addition to my collection. Send it any time: if it is a bit soft I can re-arrange it and straighten it out easily. I shall only be gone a few days on the St. L. trip.

Faithfully yours,

Louis Agassiz Fuertes

LOUIS AGASSIZ FUERTES
CORNELL HEIGHTS, ITHACA, N.Y.

Jan 5, 1917

Dear George—

The turkey came this morning. It is a beautiful skin, and a most welcome addition to my collection: the only representation of the genus I have, except for a tail & wing from Virginia. I certainly am most grateful to you, and I can see that you have put a lot of time and labor on it.

It arrived perfectly, and I have attached your card to it as a label. It gives the whole history of the specimen, and adds to its value by its personal quality—as well as for the facts it bears.

We've only had one real cold spell so far, and are just now enjoying (!) our January rain; raw, blustery and unpleasant to a degree.

Mr. Robinson has been down to the Point several times this fall, & has shot quite a few pheasants and ducks there. The owner, Mr. Liverich, died during the fall, & Mr. R. bought the whole property within the crossroads except Miss Boyer's little square, so he is now our landlord, & really owns a fine piece of property. We don't know how he will develop it, & don't care very much, for probably we shan't want to go down very many more years, & he probably won't do much for a year or two, as it cost him quite a lot of money.

When are you going to send some more pictures

for me to see? I'll be glad to criticize anything you send.

I'm deep in the little mammals now: they're great fun to do.

Best New Year's greetings to you & yours, and renewed thanks for the splendid turkey. She's a real prize to me.

Faithfully yours,

Louis A. Fuertes

I knew from an earlier letter—that of December 2, 1916—that Fuertes had no Wild Turkey in his collection, but his expression of gratitude for the specimen that I sent him was so sincere that I was made very happy by it. Fuertes was right: I *had* "put a lot of time and labor" on that truly magnificent bird. How proud I was when I thought of the label it would bear—the very card on which I had written the name of the neighbor who had shot it, the date and place of capture, and the fact that it had had a big handful of chestnuts in its crop. Blight had not yet robbed the beautiful West Virginia mountains of their chestnuts in the year 1916. I must have put my own name or initials somewhere on the card,* but I doubt that I

*The turkey specimen is still in the Fuertes Memorial Collection at Cornell University. When I last handled it myself, in the fall of 1946, its only label was the card on which I had written the pertinent data. Someone who knew nothing of the specimen's history has, since 1946, replaced the card with the formal label used throughout the rest of the collection.

said a word about the fact that I had eaten the chestnuts; they were delicious.

When would I be sending "some more pictures" for Fuertes to see? As soon as I could get them packaged for mailing. I had a Belted Kingfisher drawing that I considered good. A recent Red-tailed Hawk drawing, life-size on rough watercolor paper, was too big for easy mailing, so I would have it photographed and would send a print.

LOUIS AGASSIZ FUERTES
CORNELL HEIGHTS, ITHACA, N.Y.

Jan. 13, 1917

Mr. George Sutton,

Bethany, W. Va.

Dear George:

Your letter was very nice to receive. I am glad you liked the animals.

Any time you have anything to send me, I shall be very glad to see it.

The turkey was a great and welcome addition to my collection and was a very fine bird.

I like your kingfisher very much. It is extremely life-like. The hawk, whether or not it is due to the failure of the shadow to show up in the photograph, seems a little flat. But I know that browns do not photograph satisfactorily.

I will send you some more paper soon. It is getting a little hard now, to get the good mounted paper, for it is made in Germany, and has not been coming in. I have a few samples of Bainbridge board and paper that I like pretty well. I will send them to you to try out. You can probably get them through a stationer at Wheeling, or at least find out the home address where they are made in New York.

Faithfully yours,

Louis A. Fuertes

The letter acknowledging receipt of my kingfisher drawing and of the small photograph of my life-size drawing of the big hawk was the first typewritten letter I had received from Fuertes. I believe he had dictated it, though there were (are) no typist's initials anywhere on it.

When I undid the package in which the two items just mentioned were returned to me, I found also a kingfisher sketch that Fuertes had made but not said a word about in his letter. The sketch was so much like mine in several important ways that I felt sure Fuertes had made it either to show how different sorts of

Fuertes drawing of a Belted Kingfisher *(Megaceryle alcyon)* sent to Sutton after Fuertes had seen a Sutton drawing of the same species. Fuertes never signed this sketch.

brush strokes would produce the illusion of different sorts of feathers or to find out *for himself* whether these strokes would work the way he thought they should. To this day I do not know whether he colored-in the whole sketch with one brush. I feel sure that the long, thin, delicate crest feathers, as shown in the sketch, were the work of a large brush with fine tip, and I suspect that the short, rounded feathers of the chin and throat were the work of the very same brush managed in a different way, but of this I am not sure. What I do know is that the sketch proclaims truly remarkable expertise. Nowhere is there the slightest evidence of rubbing-out or fussing-over. I continued to expect some word as to why Fuertes had sent the sketch to me, but no such word ever came.

LOUIS AGASSIZ FUERTES
CORNELL HEIGHTS, ITHACA, N.Y.

March 1, 1917

Mr. George M. Sutton,

Bethany, W. Va.

Dear George:

I got the little pictures nearly two weeks ago and have been prevented from acknowledging them partly through a rush of closing up the first half of the animal series and absence from home on a combination business and vacation trip. They are very nice in the main. The barn swallow is extremely good. You have entirely conquered the difficulty of shadow on the under side with a tiny exception which I have corrected and which you will appreciate, I think, when I return, which will be very soon. The little dark band across the upper breast was about twice too heavy in tone where it crossed the light-struck surface. I have brought this up to the proper tone and you will see at once that it brings the whole picture into very great solidity by virtue of this single change. You must remember that black, especially when glossy, is susceptible of going very high in the tone-scale. There will probably never come a time when this proposition of color plus shadow is one of entire simplicity. I am very sure that I have to fight it all the time.

I am awfully busy catching up now, for which you will have to condone this dictated letter and also its brevity.

I would like to see your owl. I only know one man or perhaps two who have genius or have given the necessary study to even approximately preserve the beauty and interest of the eyes of the birds of prey. As a whole I like your little studies very much. Nothing is more difficult than that brilliant clear blue of the blue bird. I think you can do better with that after awhile. It is extremely subtle and difficult.

I took another liberty with the barn swallow in slightly heightening the high light on the top of his head and forehead, which I think was quite as necessary as the other alteration I made. I think the goldfinch is very pretty. You do the faces of these little finches extremely well. The whippoorwill has, in spite of its roughness of texture a very true and authentic look, that soft woodsy quality, in spite of its not having the full details. The cow bird has much of interest. It is a good bird and its general color is very true. I have none in my series nor do I remember any that is so distinctly and conspicuously streaked underneath. With my series the streaking is confined to a very narrow shaft streak which is in most cases almost a negative quantity. There is a peculiar shine on the under feathers which almost falls into streaks, but if that is what you had in mind, it is not successfully indicated in your study. The accessories are very nice; it is quite true that I have no young birds in my series and this extra streaking may be a characteristic of the first plumage, as is the buffy tinge of the throat.

I will return these studies in a few days, the first time I get to packing up. I think you have every reason to feel encouraged with the material you have sent and I think I shall be able to send you with the returning sketches, some more of the paper you like. I do not know the address of the Bainbridge company so I do not know how to get more of that paper except through a dealer.

Faithfully yours,

Louis Agassiz Fuertes

In February of 1917 I had sent Fuertes a package containing drawings of a Barn Swallow, a male Eastern Bluebird, a Whip-poor-will, an American Goldfinch, and a Brown-headed Cowbird. The last-named must have been of a female or young bird, for what Fuertes had to say about its shortcomings mentioned streaking on the "under-feathers" and a "buffy tinge on the throat." I suspect that the picture was of a young bird in juvenal plumage that I had collected at Pughtown in the summer of 1916.

The letter made abundantly clear that I still had a long way to go as a bird painter. The touching-up that Fuertes had given my Barn Swallow improved it greatly. He had lightened the dark chest band in such a way as to show just where the light struck it, and he had heightened the highlight on crown and forehead so as to show that those parts were receiving light from the same direction as that which fell on the chest. The question that bothered me sorely was this: "Why had I myself not seen what was wrong with the picture without having to be told?"

Fuertes obviously felt that my cowbird drawing left much to be desired. And the blue of the bluebird was not "brilliant" or "clear" enough. But he had been pleased with the goldfinch and Whip-poor-will; he had said in so many words that I did "the faces of these little finches extremely well"; and his summary statement toward the end of the letter was, as always, optimistic: "I think you have every reason to feel encouraged with the material you have sent . . . ."

The owl mentioned in the third paragraph of his

letter was a young Great Horn that I had for several weeks in the spring of 1917. The big fuzzy baby became the talk of the neighborhood, for it swallowed mice and half-grown rats whole, without making the slightest attempt to tear them up. Shortly before it became mature enough for return to the wild it grabbed my right hand with one powerful foot and I could not extricate myself without the help of a neighbor. There seemed to be no way in which the fingers of my left hand could pull all four claws free at the same time; so my neighbor pulled two claws free and held them while I pulled free the other two. While all this extricating was going on, the owl's huge eyes blinked amiably. "What's going on down there where my feet are?" they seemed to ask.

Sheldrake, Aug 3, '17–

My dear George—

I started a letter to you early in May! But I never got it finished, as I was rushed to death all during the spring & June, & since we came down here I've been trying—as I did last summer, to clean up the small loose ends of the spring's schedule, besides having guests here almost all the time. So I haven't really had time to give your pictures the criticism their merit deserves. I think they are as a lot much better, and some are really very excellent bird pictures, the wrens especially. I am returning them today, with real sincere apologies for having kept you so long waiting for them.

The Sharp-shin [Sharp-shinned Hawk] is criticizable as you have yourself sized it up—principally in the disproportion of the kinglet and the old fault of giving full value to the dark streaks where they come in full light. If you take an eraser & rub up the forehead, crown, breast-fluff & top angle of wing, it will show you what I mean. The breast has a translucent look—probably due to the fact that you have darkened the shaded white without correspondingly [having] darkened the shaft streaks. These hawks are so variable, though, that I hesitate to go too far in criticizing. Didn't the pale-skinned brow shield show over the eye?

The sapsucker is good, only the foot doesn't seem to encircle the branch, but to be laid on a differently pitched surface, in plan

Also, the whole light rump would go down into quite deep shadow, unless you put some reflecting surface below the bird.

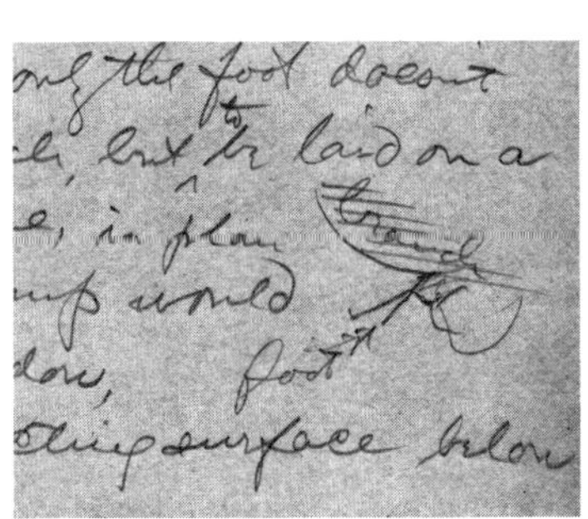
the foot doesn't
but to laid on a
in plan
would
foot
surface below

The Kingbird is nice. It's a bit full in the neck and flat in the chest for a kingbird, & the bill too thick &

arched. These big flycatchers have an almost straight culmen, sharply hooked at the end, & bristles not only over the nasal openings, but all along the gape. The tail doesn't seem convincingly copied: it would have been better to relieve the whole white tip against a darker tone than to try to alternate on the two sides. The Cowbirds are not so good as the rest, to my mind. They are difficult, & must be carefully studied. They vary & change in the changing light as the sun gets up or overhead, and should be studied intently, & all values studied relatively & digested before being painted. Then fly at it, & do it "all in one." The brown is much too yellow, instead of being filled with cold blue as it really is. (Prussian blue would ruin it, but French, New, or cobalt blue would greatly improve it. Burnt sienna, lamp-black and cobalt blue about do it; try it out and see.) The ♀ is much too pale & has the same transparent appearance due to lack of shade & light on lower & upper sides. The birds in the dove & kildee are both nice, but the landscape doesn't hang to them. Isn't the eye space of the dove a more livid light green? My memory says it is. The Ark[*ansas*]. *Kingbird is very nice. My only real criticism is that his foreshoulder & the upper surface of the wing are too "local" in color, instead of catching light, as it should, to pull that area out to the surface, it looks dished in instead of arched out.*

The yellowthroat is right good, as is the redstart. You tackled a hard proposition, & did well with it. More shade would come under the wing, & shade over bright orange is a hard thing to get convincingly in watercolor.

The winter wren is one of the very best, and exceedingly good. The little scene is nice, & sufficiently suggestive. It is hard not to make the bird look enormous. You've done very well with it. I think the bill is a bit long, dark, & somewhat turned down, for a winter wren.

The house wren is the best of the lot, and in itself I shall make no criticism whatever. You would have done well to get a chunk of bark & copied it just as lovingly as you did the bird. I think the attitude beyond criticism, and congratulate you on it as a piece of painting.

So much, then, for the pictures, which go back to you with this letter.

We have been down since June 26, & have had the same kind of a time you are familiar with. Swallows are packing for their flight. Great Blues are numerous: 4 came along recently & one staid on my big oak for a few minutes. No interesting shore birds yet.

I have 3 tame skunklets that we operated on for aromitis, successfully. You'd be crazy about 'em, as we are. Doves abound on the point, & night herons & screech owls, also. Send me more pictures when you can: I'll try to be less dilatory.

I've finished the "small mammals," which will be out next January, & am now collecting material for a series of dogs: not as much to my notion as wild things, but not to be despised in war times, when my field gets, apparently, very flat.

My regards please, to your father and mother. All the pointers want to be remembered to you.

Yours as ever,

Louis A. Fuertes

Partly because of what it had so clearly stated about some of my work's defects, the above letter had spurred me on. I must have resolved that I would show Fuertes that I could do a proper cowbird, for among the drawings that I had sent to him in the early summer of 1917 was one of a male and female Brown-headed Cowbird perched on a budding branch. The other ten were of a Sharp-shinned Hawk with a kinglet in its talons, a Yellow-bellied Sapsucker clinging to a maple bough, an Eastern Kingbird with widespread tail, a Mourning Dove with hillside background, a Killdeer at the water's edge, an Arkansas Kingbird (the species known today as the Western Kingbird), a Common Yellowthroat, a male American Redstart, a Winter Wren in a thicket, and a House Wren on a piece of bark. I had done all of these in a kind of frenzy between early March and the end of April. Fuertes had "started a letter" of acknowledgment in early May but had been so rushed by commitments that he had never finished it.

The letter of acknowledgment that he *did* write on August 3 was five pages long and in pencil. It was a very exciting letter. Like his drawings, it was full of life. As I read it I could see him sitting there in his studio at Sheldrake Point, with all of my drawings spread out before him, hard at work *writing me.* "I think they are as a lot much better, and some are really very excellent bird pictures, the wrens especially." These were his words, and they were high praise indeed.

On page 2 of his letter were three drawings, one of a sapsucker's foot clutching bark, one showing the over-all shape of a kingbird, and one of a kingbird's bill and rictal bristles. The diagram of the sapsucker's foot I did not understand: the words *branch* and *foot* should have made clear how the toes were encircling the branch, but I could never be sure which part of the drawing was the toes. The very fact that the sketch puzzled me led me to study it the more closely, however, and also to study the sapsucker's foot. What a wonderfully constructed part of the bird that foot was!

What the letter said about my Eastern Kingbird's tail reminded me in no uncertain terms that when I had drawn those twelve white-tipped feathers I had fabricated. I had put on the paper not what I had actually seen but something aimed at showing that I knew how each feather was built—with its stiff middle shaft and the webbing of barbs at each side. I squirm a bit today as I recall how I had drawn those white tips, each with light falling on one side of the shaft, shadow on the other, from imagination!

2

correspondingly darkened the shaft streaks.
These hawks are so variable, though, that
I hesitate to go too far in criticizing. Didn't
the pale-skinned brow shield show over the eye?
The Sapsucker is good, only the foot doesn't
seem to encircle the branch, but to laid on a
differently pitched surface, in plan

Also, the whole light rump would
go down into quite deep shadow,
unless you put some reflecting surface below
the bird.
The Kingbird is nice. It's a bit full in the neck
+ flat in the chest for a Kingbird, + the bill too
thick + arched. These big flycatchers have
an almost straight culmen, sharply
hooked at the end, + bristles not only over
the nasal openings, but all along the
gape

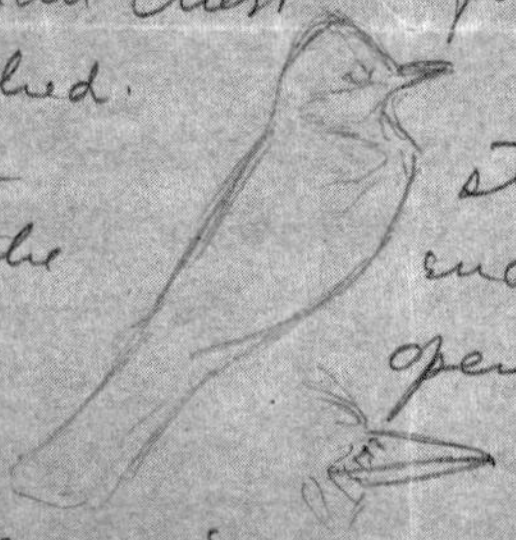

The tail doesn't seem convincingly copied; it
would have been better to relieve the whole
white tip against a darker tone than to
try to alternate on the two sides.
The Cowbirds are not so good as the rest, to
my mind. They are difficult, + must be
carefully studied. They vary + change in
the changing light as the sun gets up or

Page from letter written by Fuertes on August 3, 1917.

Over and over I read what Fuertes had to say about my House Wren. He had called it "the best of the lot." The bird itself obviously had pleased him. But what he said about the bark stuck in my craw: "You would have done well to get a chunk of bark and copied it just as lovingly as you did the bird." How many, many times I have recalled that sentence as I have struggled with the problem of making ice look cold, water wet, moss soft, petals translucent! And I have passed the word along, too. As younger men have asked me to comment on their work, I have quoted the Fuertes sentence, every word of it, verbatim.

The three young skunks that Fuertes had operated on "for aromitis," and that I would indeed have been "crazy about," were named Hyacinth, Sachet, and Violet.

LOUIS AGASSIZ FUERTES
CORNELL HEIGHTS, ITHACA, N.Y.

Apr. 29, 1918

My dear George—

Your letter came yesterday, & this A.M. the package of drawings was delivered also.

The birds are bully—far the best you've done. I can't just now go over them for you in detail, but about every one seems to have triumphed over at least one of your stumbling blocks. It's a case, of course, of developing analytical eyesight, & the only way is to do as you have done—work, work & then work.

Your greatest lack is the correlating of your bird—or "close-up," with the distance. The color-perspective isn't adjusted as it should be. To reduce this to a simple process, take a stuffed bird outdoors, and get it against a background suitable for it, with the light in front of you. Then cut a square or rectangular hole—say 5 x 7", in a fairly large cardboard—(dark color preferred, for this purpose). This is just to isolate your picture, & prevent unconscious accommodation to the whole landscape. It will enable you to see how gray, blue, or violet everything past the immediate foreground becomes. Our visual accommodation is so unconscious that without some device like this we can scarcely "believe our eyes"—or rather, hardly know what they actually see. Yet the result, when carefully studied & reproduced, constitutes only truth, by the common names of "atmosphere," "depth," etc. etc. Your landscape colors therefore must be kept in subordination. We get practically—yes, actually, no pure local color beyond the first few feet—it is always & invariably tempered by the intervening light and air—it is blued or violeted, and particularly lightened. Shadows become filled with colors—air and reflected or absorbed color from surroundings. Local color (by which I mean the actual color of the object itself, irrespective of modifying influences) persists longest, as distance is given, in the bright light. Look at a red barn, & see how little red, comparatively, goes into the composition of the *shaded side, compared with that in the lighted side. You will find the light side is oranged, in sunlight, & the shady side is full of violet. By no means are they the same red—one light & the other darker.*

I'm doing my best to fulfil a long & arduous contract—on dogs! I'm sick of it & wish it was over. It will be some day, I hope. So it may be some time before I can critically examine your last set, and return them with the criticism they merit. But you can go on, in the knowledge that these are much *your best, and show merits you haven't before achieved.*

Faithfully yours,

Louis Agassiz Fuertes

The above letter from Ithaca was so encouraging that I resolved to tackle birds that I knew would be difficult, especially species with iridescent plumage. I would make some real pictures, too, scenes with birds in them, and portraits of larger birds. Nearly all of the drawings I had thus far sent for criticism had been of little birds perched on twigs, drawings that were all foreground and no background, not even sky.

There was another reason for my eagerness to expand and improve. In the fall of 1917, at the Carnegie Museum in Pittsburgh, I had become personally acquainted with W. E. Clyde Todd, the Curator of

Birds. To my surprise I had found that this important man knew something about me. He had read my little articles in *The Oölogist* and had seen some of my drawings. He was busy with several projects and needed help. If I could be in Pittsburgh during the following summer he would arrange for my employment at the museum.

Through that fall and winter I attended college, mopped the floors and tidied the rooms of Oglebay Hall, and drew birds when I could. As a rule my model was a prepared skin. Not one drawing did I do that season directly from a living bird. Working at one of the high, stone-topped tables in the Biology Laboratory, I did first a Green Heron, the species the country folk thereabouts called the Shitepoke. The specimen's long, dark crest feathers had a lovely bronzy-green shine and a slightly dusty look, a sort of bloom. The bright colors shifted as I moved the specimen from one position to another, convincing me that the only safe procedure would be putting the skin in a certain place, leaving it right there, and painting what I saw from one angle only. I had to do from memory the colors of bill and feet, for as these featherless parts had dried, their colors had faded badly. For a perch I fetched a willow snag from Buffalo Creek. Willows and creek were my heron's background. I was immensely proud of the picture. It was a West Virginia scene with a Shitepoke in it.

The next drawing that I did was of a Common Grackle, a species then known as the Bronzed Grackle. For this drawing I had a freshly shot specimen as my model. Here the whole bird was glossy. The colors were brightest on the head and neck, but even the dullest of the short body feathers were shiny throughout their distal half. I studied the dead bird for a long while before starting to draw it. Then I propped and pinned it up in a more or less life-like position, penciled in an outline, and forged ahead. How gratifying it was to see an authentic shininess come to pass on the paper before me! The portrayal of iridescence demanded a kind of sacrifice. I could not put down all the lovely color that I knew was there. I had to be content with what I could see from one angle, and that angle only. The grackle's background was a recently plowed field.

The package of ten pictures mailed to Fuertes on April 20, 1918 included, in addition to the two just discussed, one of a pair of Tree Swallows with pale, horizonless background; one of a Blue Jay perched on a spruce branch; and one of a Whip-poor-will crouching among leaves. Concerning the last I made this not very grammatical statement in my letter of transmittal: "The leaves in the Whip-poor-will picture seem like clay rather than crisp like leaves are."

When Fuertes called my ten birds "bully—far the best you've done," he didn't underline the word *birds* in that rousing first sentence of his second paragraph. Perhaps he should have, for the latter part of the letter made clear that in none of these drawings had I correlated my "bird—or 'close-up,' with the distance." I had, in other words, been so intent on making the birds themselves acceptable that I had failed to consider them as part of what the human eye sees when it looks at a landscape. My backgrounds had shown stream-edge, willow trees, a hillside, a plowed field, what not, all with colors that I knew leaves and grass and soil had. But I had paid no attention to atmosphere, to that part of our world which hangs its veil between us and what we see. If this atmosphere is not dust-laden or smoke-filled, its veil is very thin. We usually see clearly that which is close to us; but the distant hills, whatever their actual color, are so veiled that they appear to be blue or purple. Arguing that they aren't really that color is futile: they look that way and must be shown that way in pictures intended to be at all representational.

Fuertes never criticized any one of these latest drawings in detail. He mailed them, presumably during the summer of 1918, but I never saw them again. They must have been lost in transit.

I should, perhaps, call attention here to an interesting fact. Never, since the first of my correspondence with Fuertes, had I been interested in doing pictures showing flocks of ducks or geese flying against the sky. It was the bird itself that had claimed my attention. Until the summer of 1918 the idea that a living bird, the creature that had all my life so fascinated me, was not really a living bird at all *unless it had a habitat* had never even entered my mind!

The summer of 1918 I spent in Pittsburgh, helping in the Bird Range at the Carnegie Museum. My interest in drawing led Mr. Todd to ask me to add color to the black-and-white lithographic plates in his office set of the "Catalogue of Birds in the British Museum." In the Carnegie Library's set of this several-volume work all the plates were in full color, so my job was copying these colors as accurately as possible. The work was tedious, for the plates to which I added color were never flat, and some of the colors were almost impossible to match. Part of my time I spent re-working old birdskins. The procedure involved wrapping the specimens in a damp cloth and leaving them in a "sweat-box" overnight to soften them up. Sometimes the re-worked specimens turned out well, but all too often they didn't. Some very old specimens simply fell to pieces.

Mr. Todd sensed that this not very creative work was a waste of my talents, so he asked me to draw maps and to assist with cataloguing specimens. Eventually he broached the matter of two illustrations for his forthcoming paper on "The Birds of the Santa Marta Region of Colombia: A Study in Altitudinal Distribution" (Todd and Carriker, 1922, *Ann. Carnegie Mus.*, 14, 611 pp.). I made two drawings that summer. They were of the Santa Marta Parakeet *(Pyrrhura viridicata)* and the Colombian Chachalaca *(Penelope colombiana)*, both believed by Mr. Todd to be "full species" new to science. The drawings were not very good, but I was proud of them for I knew that they were to illustrate a major ornithological work.

Other drawings that I was making in those days and that might well have been mentioned in a letter to Fuertes, but weren't, were of birds found commonly in southwestern Pennsylvania. These I made for

Charles B. Horton, a life insurance agent who let me pay premiums for two policies with bird pictures rather than money. After Mr. Horton's death, the forty-five drawings were purchased by Dr. Frank W. Preston of Butler, Pennsylvania, and presented by him to the Audubon Society of Western Pennsylvania.

The times were difficult, to say the least, in the fall of 1918. Bethany College had become a unit of the Students' Army Training Corps. I donned a uniform. As president of the senior class, I had certain extracurricular duties. I was expecting to receive my Bachelor of Science degree the following June. Because of the community's manpower shortage I was given the job of bottling and delivering the college dairy's milk, work that got me up in the middle of the night. My father was not re-hired by the college and found work in Akron, Ohio. The epidemic of Spanish influenza put my mother and two sisters to bed, all at the same time. The outspoken objections that twelve other students and myself raised to the militarization of the college led to our dismissal. But when the armistice was announced in November, I was so relieved that I felt almost nothing but joy.

The following letter from Fuertes was, in its way, the most beautiful I had ever received from anyone, for on its second page were three exquisite pencil drawings—one of a Snowy Owl, one of a deer's head, and one of a Great Horned Owl holding a squirrel in its foot. I have looked at these drawings again and again, each time deriving inspiration from them. And each time marveling. The hand that had drawn them was the hand of a genius. And that genius had gone out of his way to guide me.

LOUIS AGASSIZ FUERTES
CORNELL HEIGHTS, ITHACA, N.Y.

Dec. 9, 1918

Dear George—

Your two letters deserve better treatment than to be so long neglected. But you may have the consolation of knowing that you are not the only one I have similarly neglected.

I worked all last winter & up to August 20 on a series of dogs for the Geographic—quite out of my ken and a "war measure," being the only surviving job I had left, among a number of very interesting prospects which died, one after another, as our participation in the war developed. I went to Canada, to a boys' camp, July 15–Sept 1, and worked up there until I finished. Sumner was with me, while Mary & her mother went down to Nantucket with some friends—you remember them—the Breckenridges—at least Mrs. B, Anne & Jimmy. Wilder, an aviator in the R.F.C., was overseas, and on July 9 was shot down & captured, with a broken leg, by the Huns. His mother had a card, and later a letter from a prison camp at Limberg, but has had no word from or of him for months.

Of course Ithaca is just a training camp town now, tho' even that is now disbanding. The Aviation, Photography and Vocational schools [are] *already demobilized and the S.A.T.C., which occupied the whole university, is being mustered out, as I suppose you are, this week & last. We expect to open again, as a University, Jan. 1.*

I had a fresh horned owl and a magnificent live Snowy Owl come in at the same time not long ago, and I was pretty busy with them for a while. When I got through drawing, painting & photographing, & the white owl finally died, refusing all food & even water, I mounted them both, & they are up on my east wall looking at me now, so:

[Snowy Owl; deer head; Great Horned Owl]

I have kept busy, of course, doing what I could to push the war work of the community, & living largely on faith that things would swing back to normal, and to tell the truth the break came before I thought it possibly could. So while we shall have a more or less turbulent period of readjustment, probably, still our national health is hardly impaired, and prosperity must come back.

Don't get discouraged with me, but let me hear occasionally, and send me your work to see when you have any to send.

Faithfully yours

Louis Agassiz Fuertes

you are, this week & last. We expect to open again as a University Jan 1.

I had a fresh horned Owl & a magnificent live Snowy Owl come in at the same time not long ago, and I was pretty busy with them for a while. When I got through drawing painting & photographing, & the white owl finally died, refusing all food & even water, I mounted them both, & they are up on my east wall looking at me now, so:

I have kept busy, of course doing what I could to push the War work of the community, & living largely on faith that things would swing back to normal, and to tell the truth the break came before I thought it possibly could. So while we shall have a more or less turbulent period of re adjustment probably, still our national wealth is hardly impaired, and prosperity must come back.

Don't get discouraged with me but let me hear occasionally & send me your work to see when you have any to send.

Faithfully yours
Louis Agassiz Fuertes

Page from letter written by Fuertes on December 9, 1918.

A letter that I wrote to Fuertes late in the winter of 1918–19 must have told him of correspondence I had been having with William Leon Dawson about drawings that Dawson wanted me to make for his forthcoming several-volume work, *The Birds of California.* I was hoping that Fuertes would tell me what to charge for such drawings, for he would know precisely how good my work might be—or how bad. I was to make only a few drawings for Dawson, among them one of a male Bobolink, a well-known species that I had never thought of as Californian; one of a flying Least Tern; and one of a Yellow-rumped Warbler.

It is barely possible that I had mentioned also correspondence that I had had with Harold H. Bailey, who was planning to bring out a sumptuous volume on the birds of Florida. I was sorely in need of advice, for many plates, most of them "composite"—i.e., showing more than one species—would be needed for the Florida opus.

In 1918 and 1919 I must have given much time to writing a lengthy paper on the birdlife of the part of West Virginia that I had come to know. This paper, titled "Annotated List of the Birds of Brooke County, West Virginia," would be published serially in *The Oölogist* in 1920. Four of my drawings would illustrate it—one of a Blue-gray Gnatcatcher perched near the end of a slender twig; one of a pair of Cerulean Warblers; one of a singing Carolina Wren; and one of a Yellow-breasted Chat. Whatever their faults, these four drawings showed that Fuertes had been helping me. The editor of *The Oölogist* would go so far as to predict that "in a few short years" I would be "at the front of American bird artists" (Barnes, R.M., Editor, 1920, *Oölogist*, 37: 11). A published statement of this sort made my young head spin. It should have said, in so many words, that without the inspiration and guidance of Louis Agassiz Fuertes I might never have been able to make a drawing worthy of reproduction anywhere!

The following reply to my letter expressed joy over the fact that my work was in demand and regret over the loss of drawings I had sent for criticism the preceding April. Fuertes himself had called those drawings "bully—far the best you've done." I hated to lose them. But I shudder today as I think of the time Fuertes must have spent trying to find them.

LOUIS AGASSIZ FUERTES
CORNELL HEIGHTS, ITHACA, N.Y.

Mar 10, 1919

Dear George—

Your letter was very welcome, and it fills me with two mixed emotions—one that you have got some work in prospect with a check attached to it, the other—not so pleasant—chagrin that the pictures you sent me so long ago have not been returned. I hardly think it likely that they got lost in the mail, still, I've been all through my things, from end to end, and find no trace of them, and I quite distinctly remember writing you and sending them back. Still it is possible, at this distance from the incident, that my good intentions may have become confused & mistaken for the fact, and that the drawings are still in the studio. So I shall continue the search, and let you know later the result of my hunt. I've been pretty busy, with things that meant the assembling of an awful lot of material, and it is still possible that they may turn up.

Regarding charges, I don't quite know how to advise you. The times are very dull, and I should not think it wise to go too high, as I have had several pieces of work revoked due to high costs of presswork etc., and I should think it would be better business to be rather modest just now, and establish if possible my capability to "deliver the goods" rather than try to make the work pay its utmost. If it is for silhouette figures of birds, with little or no setting but the perch they are on, I should be inclined to do them very reasonably (especially as they are a first order) and make every effort to please your patron with the quality of the work. I make it a practice, in my own work, to do such pictures when they are for educational work of any kind particularly, for as little as I can get out of it without a loss—say a maximum of $5.00 a figure. My Bird Lore plates I do for about 3.00 a figure—even less when there are quite a number on the plate. I make my big charges on large pictures for individuals who have lots of money and want to spend it on pictures. These aren't easy to find, and they are pretty particular, too, so I don't do many *such in a year. I think my Geographic magazine contracts have been the steadiest and best earners I've had for the last few years, but probably such work only comes along after a pretty longish process of building up, & has to be carefully maintained to continue profitable to the client. In your place now I should be inclined not to exceed the figure given, but should later, if this work goes well and meets with favor, build up my price as my experience increases my value to my client.*

I congratulate you upon having found an opening, and am sure I need not tell you that the way to use it is to do your very best and keep your star high. No outside critic can be as valuable to you as you yourself, provided you keep yourself honest with yourself, & are not afraid to see the weak places in your work and labor to cure them.

Faithfully yours,

Louis A. Fuertes

Life speeded up considerably for me in 1919, for in June of that year I went to Pittsburgh to live. I had been hired by the Carnegie Museum, where I was to assist W. E. Clyde Todd, the Curator of Birds. During my first few weeks at the museum I spent most of my time remaking birdskins, arranging parts of the research collection, and helping with cataloguing; but almost every day I talked with my chief about a trip that would take him down the whole of the Labrador coast, from Battle Harbor to Cape Chidley, the following summer. He had been studying the birdlife of the vast Labrador Peninsula literally for decades. This planned-for trip would give him information that he sorely needed. He wanted me to go along as his assistant.

The expedition was not a resounding success, but all four participants returned in the fall of 1920, and the 45-foot motor yawl on which we had sailed, the *Northern Messenger,* was returned in good order to her owner, the Grenfell Mission. I had made a point of sketching, either in pencil or in watercolor, most of the bird species I had collected on that memorable trip. Among the handsomest of these was the Glaucous Gull. A large drawing of this species that I had made after returning to Pittsburgh was displayed, along with the work of several other American bird artists, at the Library of Congress in Washington, D.C., in November, in connection with the annual meeting of the American Ornithologists' Union.

I attended that meeting and made a point, of course, of looking at the bird drawings on display. Fuertes himself made the rounds of the exhibit with me, commenting—as we went from picture to picture—on composition, shadow-color, "facial expression," the failure of some drawings to measure up, the success of others. He made clear to me that while the "local colors" of my Glaucous Gull were correct enough, the "reflected color and light, on the shaded and lighted sides respectively" had not been studied-out with sufficient thoroughness. Would I never, *never* be able to see these reflected colors and record them properly?

The following letter, which had not a word to say about our Labrador expedition or about a colorplate of mine (showing the heads of eight South American quails) that had appeared in the April issue of *The Auk* (1920, Vol. 37, opp. p. 189), mentioned Major Allan Brooks, a bird painter whose work was in considerable demand at the moment. Brooks had been making many colorplates for Dawson's *Birds of California*. I had written to him, feeling that what he might have to say about the opus and its author would be of value to me.

I wish I could remember which of my drawings I had sent for Fuertes to criticize. I was glad that he, Brooks, and others had liked the "young redtail." Not for the life of me can I recall what the "little birds" were.

LOUIS AGASSIZ FUERTES
CORNELL HEIGHTS, ITHACA, N.Y.

Dec 3, '20

Dear George—

Your drawings came, were much admired by us all, including Maj. [*Allan*] *Brooks who is with me,—*

and go back to you today. I wish I had time to go over them in detail. In general, though, they bear out the same criticism I gave you in Washington, namely a sacrifice of reflected color & light, on the shaded & lighted sides respectively, in favor of what you know to be the actual local color of the part in question.

The young red-tail [Red-tailed Hawk] is very spirited and true,—wild with a real hawky wildness. Your little birds are very nice.

I should advise your using a soft pencil to draw field sketches of positions etc., & practice a free sweep of line, as against the pencil-pinching tenseness of line these small studies show. Try to feel the curves & surfaces, and sweep them in freely instead of engraving them. Get my point?

Maj. Brooks showed me your letter. I don't know how he has answered you. I think you are quite safe with Dawson, but you should make your bargain with him on a cash-for-each-plate basis. I have so much ahead that I do not think I will do any more (I've only done one!), as I don't know when I could promise anything. I think D. is all right. He doesn't inspire one with great confidence, but on the above basis you can't lose.

Hastily,

L. A. F.

In Pittsburgh I lived as inexpensively as possible, for, having decided that my parents and two sisters needed a real home somewhere, I was helping to pay for a house and small acreage along Buffalo Creek in Bethany. My Pittsburgh room—on Forbes Street, third-floor back, and only a few blocks from the museum—had very poor light, so I did all my drawing after hours at the museum. Remi Santens, the museum's chief taxidermist, saw some of my pictures and decided that I should try painting a background for a habitat group. The first such background that I tackled was for a group showing a Great Horned Owl about to capture a skunk. The painting showed sky and colorful woods, a fall scene. The group attracted a good deal of attention, for viewers could not help wondering who would win. The owl was obviously big and powerful, but the skunk was, as everyone knew, anything but defenseless. Under the circumstances people paid little attention to setting or background. The confrontation of owl and skunk was what mattered.

The background painting was in oils. This was a medium I had never before even tried. It fascinated me. I obtained an easel, ordered a stretched canvas or two, and did some paintings, notably one of a Great White Heron that was part of a display at the Carnegie Museum in the fall of 1924 and of a similar display at the Los Angeles County Museum in California in April of 1926. This oil was reproduced as a halftone in *The Condor,* the bimonthly journal of the

Cooper Ornithological Society (1926, *Condor*, Vol. 28, p. 201).

I have gone into some detail about the Great White Heron painting chiefly because a Fuertes watercolor showing two Prairie Falcons, one with a quail it had captured, the other in the distance, flying, was painted especially for the show at the Los Angeles County Museum. Harry Harris, who reported on that show, had this to say about Fuertes: "It is indeed fortunate for ornithology that this artist has held to absolute freedom in developing his own method of expressing his fine appreciation and feeling for bird beauty, thus preventing his approach from being prejudiced or hampered by academic training. That he was early saved to science by the acumen and sound judgment of his mentors is cause for the thanks of all bird lovers, for here is a painter after the hearts of them all. His Prairie Falcon, graciously done especially for this exhibition, is so instinct with truth, character, action and *expression* that the very acme of avian portraiture has been reached in it. A black and white reproduction can not of course do justice to the original" (Harris, 1926, *Condor*, 28: 193, 202).

This same Fuertes had been my guide and mentor since the spring of 1915. To him I had sent for criticism the three drawings to which he referred in the following letter. I found it hard to believe that this great man had expressed a hope that I would send him "more to look at." Would he never tire of helping me? Would his belief in me never waver?

Sheldrake, N.Y.
July 23, '21

Dear George—

You must never think you are presuming or otherwise troubling me by sending me work to look at,—especially when it is as nice as these last three studies. They are excellent, and each one has its own particular charm and personality. The kinglet is first class, and a true kinglet in every line. The delicate little manner of standing on its tiny legs is very nicely caught. You can't do too much of this work of studying & noting & drawing from life: there's no other source of the kind of information you require, and you'll find you learn a lot every time you draw from a live bird.

The Song Sparrow is also a very convincing little drawing, full of Song Sparrow style and authenticness.

The Waterthrush is excellent, too, but less remarkable, I think, perhaps because left a little less finished. It is a good attitude, and it is very nice and solid in values.

I congratulate you on these, and hope you will send me more to look at.

My regards to Mr. Todd.

Sincerely,

Louis A. Fuertes

At the Carnegie Museum I found myself part of the great, wide, wonderful world. Every day I handled specimens from various countries of South America, chiefly Colombia, Venezuela, and Bolivia. A collection of birdskins from the Province of Hopeh in China, I was asked to reshape. The museum's collection of birds' eggs, a small but important one, needed a curator's attention. There was no telling what important person might come by for a visit. On one occasion the museum's director asked me to give the famous Norwegian arctic explorer, Fridtjof Nansen, and his lovely daughter a guided tour of the exhibition halls.

The Riess [sic] referred to in the following letter was Rev. Jacob A. Reis, who lived for some time at Lolodorf, Cameroons, western Africa, built up a fine collection of birds, mammals, and insects while there, and sold these to the museum. The "Mr. Good" was the Rev. A. I. Good, also presumably a missionary, who lived at Metet, Cameroons. My friend Kenneth C. Parkes, Curator of Birds at the Carnegie Museum, informs me that the Reis bird collection is now the "largest single African collection" of birds at the museum; that the museum has only a few birds collected by Good; and that Good presented to the museum eleven photographs of African mammals (four of gorilla, two of chimpanzee, three of leopard, and two of pangolin). These were prints, not negatives, and Good did not sell them to the museum; he gave them. When Mr. Reis showed me prints that he had with him of the four gorilla photographs, I knew at once that they would be of great interest to Fuertes.

I cannot remember whether Mr. Reis actually gave me the prints; I believe he merely lent them. If he did so, I returned them to him after Fuertes had had a look at them.

LOUIS AGASSIZ FUERTES
CORNELL HEIGHTS, ITHACA, N.Y.

[*undated, about Nov.1, 1921*]

Dear George—

It was mighty good of you to send me the Gorilla prints: Mr. Riess [*sic*] *had told me of them, and I at once wrote Mr. Good, and your pictures came in the same mail with a letter from Good promising to send me prints later, when he could get time to make them. Has he the negatives? I should have supposed he'd have sold them with the skin etc. to the Museum. They are certainly wonderful, & give a lot of very interesting data concerning the Gorilla's living appearance. How different in every respect from the Chimp and Orang!*

I'll bring them to Philada. next week & hand them to you: I want to make some "transcriptions" from them as to details.

Thank you ever so much for sending them. See you in a few days,

Sincerely yours

Louis A. Fuertes

Hardly a day passed at the Carnegie Museum without my hearing from Mr. Todd about plans for his "North Country work." He was negotiating with collectors of specimens in South America and preparing detailed taxonomic reports on collections already received from that part of the world, but all the while he was dreaming of a project he had started a full quarter-century before when, after "skirting the shores of the Gulf of St. Lawrence" he had spent "a month or two on the east Labrador coast" (Todd, 1963, *Birds of the Labrador Peninsula and Adjacent Areas,* Univ. Toronto Press, p. vii). He was a man of stubborn, almost grim, resolve. The heavens might fall, but he would climb over whatever ruins there might be of pearly gates and golden pavement to that New World "North Country" on which he'd promised to report.

A part of the Labrador Peninsula that he wanted to visit again was James Bay. A wealthy man named John B. Semple offered to finance an expedition to that part of Canada to obtain specimens of Blue Geese for a habitat group at the museum. Mr. Todd made clear that if such an expedition came to pass, he would expect me to go along as his assistant.

Naturally I was excited. Naturally I wrote Fuertes when I learned that I'd actually be going to James Bay. In my enthusiasm I may have gone a bit too far when I asked about specimens that I might obtain for his wonderful collection. I knew that Mr. Todd had agreed with me that gaps in the Fuertes collection should be filled if possible. Mr. Todd knew very well how I felt about Fuertes. Be all this as it might have been, Fuertes reminded me that I would, on that trip to James Bay, have an employer, and that that "employer's interest" was what I should be working toward first of all. Of course Fuertes was right.

Our "Blue Goose Expedition" did not return to the United States until October 11. The American Ornithologists' Union met in Cambridge, Massachusetts from October 8 to 12. I feel quite sure that Mr. Todd did not attend the meeting. I know that I didn't.

Sheldrake

Aug 5, '23

Dear George—

I was mighty glad to hear from you, & to learn that you've got such a fine trip ahead. I congratulate you, and wish you the best of success.

I don't know exactly what to expect from James Bay: it's a big place & I don't know much about it. I should be much interested to hear, when you come back, what you find there. Don't bother to have me in mind, for I have no idea, as I say, what you will strike, & in any case your time should go to your own & your employer's interest. Anything that I might want would in all probability be very desirable to your own uses, & I can perhaps borrow it, later, if the need arises. I am pretty well stocked with E[*astern*]. *N*[*orth*].

Am[erican]. shorebirds—both spring & fall, as I have concentrated thereon.

We are all well—as you see—at Sheldrake again. I bo't a used Dodge this summer, which frees us somewhat—at least makes us feel we can get away—so we seldom care to.

I've delivered 13 Mass[achusetts]. Plates only: have 2 more done & am carrying it on this summer—not very hard, but still progressing. It had a year-long lapse in favor of pre-contracts, but now should sail along, at least abreast of the author. Shall we see you in Cambridge at A.O.U.?

Sincerely

L.A.F.

One of the most important achievements of our "Blue Goose Expedition" was discovery of a population of Le Conte's Sparrows in the prairielike area just south of the vast James Bay tidal flats. This was a species with which I had become acquainted on its wintering grounds in Texas. I had long considered it one of the most delicately beautiful of North American birds. In the following letter from Fuertes he called it "the mousiest bird" known to him. What a wonderful way to describe it! His two drawings of it, in one of his early composite plates in *Bird-Lore* (1910, Vol. 12, frontispiece), were exquisite.

The "falcon matter" discussed in the letter's long second paragraph was of special interest to me. I had been told by somebody, one of our Indian guides probably, of a Peregrine that had been seen to strike a flying loon, killing both the loon and itself. I had passed this somewhat wild tale along to Fuertes, to see what he would think of it.

Many a wild tale is told in the North Country. Everyone who travels that way knows this to be true. Such tales take the place of movie shows and newspapers. One of the wildest heard by me from a white man was about a trapper who, suddenly attacked by appendicitis, froze the lower right side of his abdomen, removed the offending part of his viscera, and survived. Hyperbole in tales told by whites in the North is to be expected. Usually it harms no one. It is quite insipid, innocuous, and unexaggerative, I may say, as compared with that in Eskimo tales.

I am allowing myself this digression, for I have often wished that Fuertes could have heard the argument I had with a fine old Aivilik Eskimo named Angoti-Marik over a tale he had told me of a woman who had rubbed the spirits the wrong way, leading them to destroy her, the igloo in which she was living, and all her personal possessions by fire. "But you know very well," I said to Angoti-Marik, "that snow doesn't burn, so how can you expect anyone to believe such a story?" The spinner of yarns, white of hair but as straight as a cleaning rod, replied: "I know that snow does not burn, but what I have just told you is what happened all the same!"

As for the Blue Goose, I had set my heart on finding that controversial bird's nest. Many ornithologists believed that it and the Snow Goose were one and the same species, a concept that is widely held today. Fuertes made an important point, however, when he called attention to the fact that no blue phase of the *Greater* Snow Goose had ever been reported. The phenomenon is puzzling, to say the least. It leads one to wonder why the "fittest" that survives in the long and supposedly demanding process of evolution should vary so much phenotypically. Why work so hard to prove yourself "the fittest" if there's simply no telling whether you'll wind up white or "blue" with white head and neck?

LOUIS AGASSIZ FUERTES
CORNELL HEIGHTS, ITHACA, N.Y.

Oct 26, 1923

My dear George—

I was mighty glad to hear from you, & to learn how successful your trip north had been. I had advance aero. rumors of you at Cambridge. Congrats on LeConte's Sparrow: it's a wonderful little bird—the mousiest bird I know.

The falcon matter is one hard to prove either way, but my belief is that they hit their prey a terrible wallop with their half open fist, adding to that bolt the impact & momentum of their stoop, & that they only use their breast bone very secondarily, if at all. I've never seen any sign that it was used, & certainly it has no special development, nor callous, & what, otherwise are those hellish feet for? I've never heard more than mythical & wildly improbable tales of a hawk killing himself in his stoop, altho' if he hit a loon weighing some ten pounds, his pound & a half— (or her two & a half) would receive a very severe jolt & might well knock her for a loop. I think most people find it very easy to introduce the mystic viewpoint when dealing with the peregrine, and you would certainly be safe, at least, in sticking to the conservative side of it. I personally do not believe the peregrine does anything radically different from what any other falcon does, & I feel pretty sure that if they hit with the body it is because they have missed with the feet first.

The Blue Goose matter is quite a puzzle, & I don't know much about it, though it seems that if it were a phase of the Greater Snow the Lesser Snow and Snow Goose of Europe would have something comparable to it—which I have not heard they have. I know of no Blue Goose that ever developed (in captivity) out of the dark dress into the pure white, further than to assume, after 3 or 4 years, a white belly and perhaps a few white feathers on the side and upper surface. Kennard has gone as far with this as anybody, but I'd ask the opinions of Lee Crandall at the Bronx, &

Ned Dearborn at the National Zoo in Washington. Both have had blue geese for many years in captivity.

My regards to Mr. Todd.

Sincerely

L.A.F.

The "Allen" referred to in the following letter was Arthur A. Allen of Cornell University, the professor under whose direction I would take my Doctor of Philosophy degree some years later. "Lloyd" was Sam Lloyd, a friend of Allen's. "Dr. Fox" was a practicing physician of Houston, Texas. The "another" was Jim, Dr. Fox's boatman. The four men were duck hunting along the coast not far from Galveston, and a harrowing time they had when "one of those unpredictable northers" struck. Allen's account of the experience appears as a chapter in *Discovery,* a book dealing with "great moments in the lives of outstanding naturalists" (Terres, J.K., Ed., 1961, Lippincott, Philadelphia and New York, pp. 26–31).

Fuertes inquired if I was busy. Indeed I was, for I had decided to leave the Carnegie Museum, to move to Harrisburg, there to become Pennsylvania's State Ornithologist and Chief of the Bureau of Research and Information for the Board of Game Commissioners. Too, I had promised Harold H. Bailey to complete as soon as possible the seventy-six full-page colorplates for his forthcoming *Birds of Florida.* The first of these was of American Flamingos, a species I had never seen in the field. It was the only picture of the lot that showed one species. The rest were "composite," each showing from three or four to eight or nine species. Not one of the plates did I send to Fuertes for criticism. There seemed to be no time for that. But as I worked on them I bore in mind what he had told me about the colors of shadows and reflected colors. In Witmer Stone's review of the book he said: "The most regrettable feature of the present work . . . is the fact that in some of the plates the blue has been given entirely too much prominence and has been used in shading the under side of white-breasted birds so that we find the White-throated Sparrow with a bright blue breast and the Willet with blue and black wings and a blue belly, while the overlay of blue on the yellow of the Prothonotary Warbler and some other yellow-breasted species renders these parts a vivid green. We can hardly imagine that the artist was given an opportunity to pass on the colors of the reproductions and whoever approved them was woefully incompetent" (W[itmer]. S[tone]., 1926, *Auk,* 43: 105).

The fact is that I, the artist, had done my best in correcting proofs, for I wanted those plates to meet with the approval of bird students the country over, and especially of Fuertes. In painting the originals, however, I had not made allowance for the tendency of engravers to "improve" colors, i.e., to make them brighter or "more attractive." In this connection I recall so vividly the statement of an engraver who was to reproduce some of my work: "Now don't you worry about a thing. What we'll do to these pictures of yours will improve them, just remember that!"

In my opinion the poorest of my drawings for Bailey's book was that of the Scissor-tailed Flycatcher. Here was a bird that I knew well, that I'd grown up with on the Texas prairies, yet crowding it into that wretched composite plate forced me to show it perching vertically. Scissortails simply do not perch that way as a rule; they perch horizontally, facing the wind, with that fine tail of theirs sticking straight out behind them. I should have arranged the six bird figures on the plate in a wholly different way.

The "Alfred" mentioned in the last sentence was, I believe, my good friend Alfred Emerson, an authority on termites, who was on the faculty of the University of Pittsburgh at the time.

201 Wyckoff Avenue
Ithaca, N.Y.

Jan 28, '24

Dear George—

Allen & Lloyd are all right. They got stranded last Saturday (19th) a few miles from Galveston, & two of the party (Dr. Fox & another) rowed back to

Galveston for supplies & help. That was on Tuesday. We got the A.P. report here at 10 A.M. Tuesday, and at 11:30 another report that the boys were safe & comfortably camped on the "Island," and would continue on their present trip. I think no further news has come in, but feel no apprehension, as they are tried and experienced trippers, and would not take any foolish chances. Fox was both exhausted & excited, apparently, when he first got in, & the first report was alarming & had us all up in the air, but was so soon followed by the more plausible and comforting message that we ceased worrying forthwith.

Busy? I presume you are. I've recently used your hen turkey to advantage as model for most of the birds in the enclosed photo of a big 30 x 56 oil I've just finished for a man in S. Carolina. I got a cock last year from Virginia—17½ lbs. and very gorgeous. Regards to Todd & to Alfred, & to yourself too—

Yours

L. A. F.

In the fall of 1924 the American Ornithologists' Union was to convene in Pittsburgh for the first time in its history. The meeting would continue from November 10 to 13. At the request of Mr. Todd, who had been assured by the Department of Fine Arts of the Carnegie Institute that they would provide an adequate gallery and help with hanging the pictures, I entered into correspondence with bird artists far and wide in the United States and Canada, asking them to lend their work for an exhibition to be opened during the meeting. The response was excellent, and the show, which continued to be viewed by the public for three weeks, was considered "the chief feature of the meeting" (Palmer, T.S., 1925, *Auk*, 42: 110).

Fuertes was, of course, among the thirty-one artists who lent their work for this exhibition. Shortly before the meeting I wrote him, inviting him and his wife to have dinner with me at a hotel not far from the museum on the evening of the show's formal opening. In the following brief letter he accepted my invitation. Seeing that dinner through to a successful conclusion was something of an adventure for me, for never in my life had I attempted to entertain in honor of so distinguished a guest. Among those seated at the table with Fuertes and me were Bayard H. Christy, editor of *The Cardinal*, the journal of the Audubon Society of the Sewickley Valley; John B. Semple, who had financed the "Blue Goose Expedition" to James Bay the year before; and five others. It was quite an occasion.

The "bulky box" mentioned in the letter was, of course, that which contained the pictures he had sent for the exhibition.

LOUIS AGASSIZ FUERTES
CORNELL HEIGHTS, ITHACA, N.Y.

Mon. A. M.

Dear George—

Your note just came. I shall be most happy to join you as you suggest, only, I am sorry to say, I shall be alone, Mrs. Fuertes being unable to come with me.

I presume my bulky box came in due season, & I hope the stuff will be of the right sort. I had no time last A.O.U. to do anything for it.

Hastily, & with many thanks,

Louis A. Fuertes

How very ironic that not until I had finished all but two of the seventy-six plates showing most of the four hundred or so species and subspecies of birds known to inhabit Florida and its adjacent waters, should I at last *see* that exciting subtropical part of the world! I had made the drawings in Pittsburgh. While doing my best to depict live-oaks, cypresses, sweet gums, mangroves, palmettos, water-hyacinth, and saw-grass acceptably I had often thought of what Fuertes had said to me about finding a piece of bark and painting it just as "lovingly" as I had the House

Wren. There I was on the third floor of the big museum building imagining what wild Florida looked like! The room's only windows were so high that if I wanted to see more than sky through them I had to get a ladder or climb onto the top of one of the metal cases that held specimens.

In the early spring of 1924 I actually went to Florida. John B. Semple, with whom I had become well acquainted on his "Blue Goose Expedition" to James Bay the year before, invited me to accompany him to Cape Sable, Flamingo, Gator Lake, Key Largo, and other places that he knew about in southernmost Florida to collect specimens for the Carnegie Museum. I was afield with him from March 10 to April 3. Our findings were reported in a paper published in the museum's *Annals.* A separate of this paper I had sent to Fuertes and in this, the last letter I was to receive from him, he thanked me for it.

The letter was characteristically warm-hearted, thoughtful, and exuberant. Fuertes was elated over the "hundred color studies" and "lot of drawings" that he had made, as well he might have been, for both separately and as a collection they are outstanding for their beauty, their dramatic quality, and their authenticity. I often pore over my boxed set of thirty-two of them and gloat. How proud I am that any fellow human being has been able to achieve such success in recording the special beauty of wild creatures!

The letter's fifth sentence—about herons—needs explanation. The heron that Fuertes here calls the "white phase of *A. wardii*" was for some time considered a "full species" and known as the Great White Heron *(Ardea occidentalis)*. It was so listed in the Fifth Edition of the American Ornithologists' Union's *Check-list of North American Birds* (1957, p. 40). Taxonomists now believe, however, that this all-white heron and the equally large but dark heron, the bird that Fuertes calls "A. wardii," a form that inhabits the southeastern United States, are morphs or color-phases of one and the same well-known, wide-ranging species, the Great Blue Heron *(Ardea herodias)*, and that the so-called "Würdeman's Heron" is an intermediate and variable bird found only in the area throughout which the ranges of the white morph and the blue overlap—*viz.*, Florida and the Florida Keys. Fuertes was, in other words, essentially in agreement with present-day thinking when he called the "case" similar to that of the Snow Goose *(Chen caerulescens)*, whose white and blue morphs were for a long time considered two distinct species.

201 Wyckoff Avenue
Ithaca, N.Y.

June 26, '27

Dear George—

Your letters are among those I found on my return recently from Abyssinia. Thank you for them and for the "Notes on Birds of Southern Florida." I am delighted to have it. I got a great thrill myself out of my brief visit at Cape Sable & Cuthbert Lakes, with the Würdeman's Heron colonies on the Keys. I don't personally, consider W's [Würdeman's] Heron of any taxonomic value—merely an uncompleted white phase of A[rdea]. wardii, of which [Ardea] occidentalis is the perfection of a white phase, = similar case of A[nser]. coerulescens.

We had a marvellous trip in Abyssinia, & among other things I got far the best lot of field studies I ever did on one trip: a hundred color studies & a lot of drawings. Many of the birds most curious & bizarre.

I've so much now to do to catch up my neglected work that I can't see which way to turn. But I must turn somewhere, so have to sign off before I write you a book on Abyssinia. I've had no chance to talk to any but the home-town folks, interested mostly in the people & the country. I think I'd talk a bird man into an early grave!

Faithfully

Louis A. Fuertes

Fuertes never said a word to me about the drawings I had made for the Florida book. He was in Africa for some time, to be sure, but almost certainly he had seen the book before going there, for it had been published in 1925. I cannot help feeling that those colorplates of mine disappointed him. Indeed, they should have, for as a whole they were not very good and some of them were shamelessly imitative of him in both concept and style.

How the true Fuertes—friendly, appreciative, enthusiastic—shone forth from the words of his last letter! He had greatly enjoyed Abyssinia. He would gladly have gone on and on about his experiences in that country had he not been obliged to get back to "neglected work." He was fairly boiling over to tell it all to "a bird man"!

I learned all too belatedly of the grade-crossing accident that killed Fuertes on August 22, 1927, near Unadilla, up-state New York. A friend in Harrisburg telephoned me, saying that he'd seen a newspaper item about it. I was so stunned, so bewildered, that I shut my office doors, locked them, and found a chair. Fuertes, the one and only Fuertes, gone! The Fuertes who had been my friend, my mentor, my maestro—gone! And at the very peak of his career!

How easy it would be to say that Fuertes is not really dead; that he lives on through his pictures; that he is indestructible; that the very force which drives me, at times relentlessly, to do the best I can at drawing birds, *is* Fuertes. I know that thoughts of this sort are a kind of romanticizing, a way of imagining light where there is nothing but darkness. I know that Fuertes is no more.

But only this morning I watched a recently fledged cowbird, fifteen feet or so from me, walking in a gravelly spot on the campus, trying to find something to eat. Its foster parent, a Cardinal, was not far away, refusing to feed it, for the time had come for the cowbird to be on its own. The cowbird looked at the Cardinal, fluttered its wings beseechingly but in vain, then looked at me, perhaps wondering why I should be standing there towering over it as I was. And the look on its face reminded me of Fuertes, of the incredible know-how that Fuertes had. That look, that special look on the little bird's face, Fuertes would have been able to put on paper. Facts of this sort about the man will abide as long as there are human beings who take the time to look, really look, at living birds and at pictures of them. What could be closer than that, I ask, to immortality?

BY GEORGE MIKSCH SUTTON

To A Young Bird Artist: Letters from Louis Agassiz Fuertes to George Miksch Sutton (Norman, 1979)
Fifty Common Birds of Oklahoma and the Southern Great Plains (Norman, 1977)
Portraits of Mexican Birds: Fifty Selected Paintings (Norman, 1975)
At a Bend in a Mexican River (New York, 1972)
High Arctic (New York, 1971)
Oklahoma Birds: Their Ecology and Distribution, with Comments on the Avifauna of the Southern Great Plains (Norman, 1967)
Iceland Summer: Adventures of a Bird Painter (Norman, 1961)
Mexican Birds: First Impressions (Norman, 1951)
Birds in the Wilderness (New York, 1936)
Eskimo Year (New York, 1934)
The Exploration of Southampton Island, Hudson Bay (Pittsburgh, 1932)
An Introduction to the Birds of Pennsylvania (Harrisburg, 1928)

ILLUSTRATED BY GEORGE MIKSCH SUTTON

A Guide to the Birds of South America (in part) (by R. Meyer de Schauensee) (Wynnewood, 1970)
The Birds of Colombia (in part) (by R. Meyer de Schauensee) (Narbeth, 1964)
The Birds of Arizona (by Allan Phillips, Joe Marshall, and Gale Monson) (Tucson, 1964)
Birds of the Labrador Peninsula and Adjacent Areas (by W. E. Clyde Todd) (Toronto, 1963)
Fundamentals of Ornithology (by Josselyn Van Tyne and Andrew J. Berger) (New York, 1959)
Georgia Birds (Norman, 1958)
A Guide to Bird Finding West of the Mississippi (by Olin Sewall Pettingill Jr.) (New York, 1953)
A Guide to Bird Finding East of the Mississippi (by Olin Sewall Pettingill Jr.) (New York, 1951)
World Book Encyclopedia (section on birds) (Chicago, 1941)
Birds of Western Pennsylvania (by W. E. C. Todd) (Pittsburgh, 1940)
Texas Bird Adventures (by Herbert Brandt) (Cleveland, 1940)
The Golden Plover and Other Birds (by A. A. Allen) (Ithaca, 1939)
Wings, Fur and Shot (by Robert B. Vale) (Harrisburg, 1936)
American Bird Biographies (by A. A. Allen) (Ithaca, 1934)
The Birds of Minnesota (in part) (by T. S. Roberts) (Minneapolis, 1932)
The Burgess Seashore Book (in part) (by Thornton Burgess) (Boston, 1929)
The Birds of Florida (by H. H. Bailey) (Baltimore, 1925)
Our Bird Friends and Foes (by William Atherton Dupuy) (Philadelphia, 1925)